Multivariate Analysis of Variance (MANOVA)

Multivariate Analysis of Variance (MANOVA)

A Practical Guide to Its Use in Scientific Decision Making

Harry R. Barker
AND
Barbara M. Barker

The University of Alabama Press

Library of Congress Cataloging in Publication Data

Barker, Harry R., 1924–
  Multivariate analysis of variance (MANOVA)

  Bibliography: p.
  Includes index.
  1. Multivariate analysis. I. Barker, Barbara M., 1936– II. Title.
  QA278.B37   1983   519.5′35   82-16122
  ISBN 0-8173-0141-0
  ISBN 0-8173-0142-9 (pbk.)

To the memory of
Daniel Ross Barker

Contents

Preface   ix

Introduction   1

1. Historical Origins of MANOVA   5
   Era of Multivariate Techniques
   Sequential Trends in Application of Multivariate
      Techniques

2. Conceptual Theory Underlying MANOVA   14
   Parallels between Univariate ANOVA and Multivariate
      MANOVA
   Factor Analysis and MANOVA
   MANOVA Tests of Statistical Significance
   Differential Sensitivity of Test Criteria Related to
      Distribution of Trace
   Assumptions Underlying ANOVA and MANOVA

3. Decision Strategies   28
   Decision Errors
   ANOVA Power Analysis
   MANOVA Power Analysis
   Bonferroni t
   Classic MANOVA Procedure
   Hummel–Sligo Procedure
   Mixed Strategy

4. Classic Research Designs   42
   Two Preliminary Issues
   Control Checklist

Origin of All Classic ANOVA Designs
Extension of t Test for Independent Groups
Extension of the t Test for Matched Pairs (Subject as
    His or Her Own Control)
Mixed Designs

5. Applications of MANOVA to Classic Research Designs   52
Preliminary Considerations
Classic Designs

6. Application of MANOVA to Univariate Designs That   96
    Involve Repeated Measures
Distinction between MANOVA Applied to Univariate and
    Multivariate Repeated-Measures Designs
Univariate Analysis of Repeated Measures
A Univariate Procedure for Analyzing Repeated-Measures
    Designs
Multivariate Analysis of Variance of Repeated-Measures
    Designs

7. Checklist for the Investigator Conducting MANOVA   102
    Research
Decision to Conduct a Study or Experiment
Selection of Dependent Variables
Selection of a MANOVA Test Criterion
Statement of Problem
Research Design
Computer Program Test
Selection of MANOVA Strategy
Hierarchy of Hypotheses
Reporting Multivariate Outcomes

Appendix. Hand-Calculated Example of One-Way   108
    (Simple Randomized) MANOVA

References   124

Index   128

Preface

The authors owe their interest in and fascination with multivariate techniques largely to the earlier influence of Sam Webb. For help with this book we acknowledge informally a very large number of graduate students and colleagues; we benefited greatly from interaction and stimulation regarding multivariate topics in the classroom and in the research arena. The participants in our MANOVA workshops helped us articulate many difficult points. We must formally thank the following colleagues, who read an early draft of the manuscript and responded with encouragement and numerous criticisms: Ajit Mukherjee, William Osterhoff, Betty Carlton, James McLean, Steve Prentice-Dunn, Robert Cruise, David Sohn, Carolyn Minder. David Lane made particularly astute suggestions that were helpful in improving the text.

We are deeply grateful to several anonymous referees who read earlier and later versions of the manuscript and offered extensive constructive criticisms and encouragement. Despite the considerable help we have received from students, colleagues, and referees, we remain in disagreement on several issues in the manuscript. We assume sole responsibility for whatever faults remain and hope they are few!

HARRY R. BARKER
BARBARA M. BARKER

Multivariate Analysis of Variance (MANOVA)

Introduction

Considering the plethora of statistics books currently available, a convincing raison d'être for any new statistics text seems altogether desirable, if not essential. The principal justification for this text is that it relates an advanced topic (MANOVA) to conventional statistical topics, thereby making it intelligible to the majority of investigators. This approach is in contrast to that taken by existing texts on MANOVA, which deal extensively only with the underlying mathematical theory.

The mathematics underlying MANOVA is complex and is most readily handled via matrix algebra, which is not well known to most researchers. Numerous good texts are available that describe the mathematics involved in MANOVA; those interested should consult Anderson (1958), Tatsuoka (1971), Timm (1975), Bock (1975), Morrison (1976), and others. In contrast to this abundance, there is an utter scarcity of works that address the topic in a practical, applied fashion, as this text is designed to do. In keeping with the "new" look in statistics (Wallis & Roberts, 1956), we focus on MANOVA as a tool that can aid the researcher in making wise decisions regarding the design of research and the analysis of data. Treatment of topics is conventional in that it links MANOVA to previously known conventional concepts, and explanations are kept at a meaningful decision level. The contents of the book, in abbreviated form, have been successfully presented to participants in numerous workshops with a wide variety of backgrounds in the physical and social sciences and mathematics. Workshop participants included full-time researchers, faculty members, and students in training. The workshops varied in length from 1 to 3 days.

The text should be useful as a supplementary text in courses involving experimental design and multivariate techniques. The minimal level of statistical knowledge assumed is a graduate-level course in analysis of variance. In addition, a superficial acquaintance with correlation techniques and factor analysis would be helpful. Ideally the reader would have background training of the sort reflected in the text *Statistical Analysis in Psychology and Education,* 5th ed. (New York: McGraw-Hill, 1981), by George A. Ferguson. The approach used in the present work should be of value to both the budding prospective scientist in training and the sophisticated, previously trained scientist. Perhaps a large group of users will be those faculty members and researchers whose statistical training was completed prior to the advent of computer programs and who wish to relate the principles of multivariate analysis of variance to their knowledge of earlier, more conventional techniques.

We leave mathematical analysis in the strict sense of derivations to texts already available; we emphasize treatments of topics at the global conceptual level as they relate directly to the decision maker. This approach enables the researcher to avoid numerous pitfalls of interpretation that frequently accompany the application of a "new" method of analysis to research data. For example, recognition that characteristics of MANOVA are conceptually related to well-known multiple regression concepts, such as the suppressor variable, allows an investigator to look beyond the dependent variable's failure to separate treatment groups, so that the researcher identifies its possibly important contribution to a synthetic variable that effectively displays a treatment effect.

In chapters 5 and 6 the reader should have the sense of looking over the authors' shoulders as they deal with actual research data in an appropriate fashion. In keeping with a practical approach, computer solutions of MANOVA have been used throughout (except for the one hand–calculated example provided for the curious in the appendix). The reader is afforded an opportunity to follow the decision-making process from statements of the research problem through selection of measurements and research design and interpretation of computer analysis, and we have highlighted contrasts between actual and ideal procedure. An unexpected advantage results from a didactic approach that links MANOVA with better-known statistical concepts. One rapidly develops a greater awareness of the need in behavioral research for measures that are reliable, valid, and of appropriate dimensionality. It is hoped that increased sensitivity toward measurement may ultimately improve the quality of research.

The book has been organized and developed so that the active researcher may turn directly to a particular chapter that meets particular needs of the moment. We view the sequential arrangement of the chapters as appropriate for the investigator first becoming acquainted with MANOVA. In conducting numerous workshops on MANOVA, we have found a sequence that seems to us most natural: the material of chapter 3 followed by the one-way MANOVA example in chapter 5. Then the materials included in chapters 1, 2, and 4 are presented. The more complicated MANOVA designs of chapter 5 follow. Because time available for workshop presentations is limited, the checklist for MANOVA in chapter 7 is hastily introduced with very brief allusion to the use of MANOVA for univariate repeated-measures designs (chapter 6).

Chapter 1 of this book presents the historical origins of MANOVA, viewed as a logical outgrowth of several previous conventional trends in research methodology. The chapter seeks a meaningful answer to the question "What is MANOVA?" in the context of more familiar statistical concepts and methods. Although the mathematics of MANOVA was known for many years, its great complexity and the large amount of hand calculation required for solution obscured its usefulness and prevented its general adoption.

Chapter 2 deals with the conceptual theory underlying MANOVA. We place considerable emphasis on the conceptual equivalence of univariate ANOVA and multivariate MANOVA. This approach allows the investigator to acquire rapidly a sense of the vital contribution that MANOVA can make in research. Other familiar, conventional concepts in statistics involving factor analysis and factor scores are discussed to highlight the essence of the MANOVA approach.

Chapter 3 gives some decision strategies for MANOVA. Several well-developed decision strategies are available to the investigator who conducts MANOVA research. These strategies are explained, and their advantages and disadvantages are examined, with attention to the distinctive role of MANOVA in research.

Chapter 4 presents some of the classic research designs and gives a simple paradigm and nomenclature for identifying and relating all research designs. It may serve as a very brief introduction to design of research for the neophyte or as a quick review of research design principles for the sophisticate. The principal value of the chapter is the standard schema and nomenclature, to which we refer throughout the book.

Chapter 5 deals with applications of MANOVA to some of the classic research designs. One of each of the classic research designs is illustrated from the initial step of formulating a research problem and selecting measures of the dependent variable(s) to interpreting an actual MANOVA computer printout. This portion of the book affords the reader an opportunity to see how the authors apply MANOVA principles from the earliest stages to the conclusion of the research process.

Chapter 6 describes the application of MANOVA to repeated-measures univariate designs. A growing trend in the analysis of univariate repeated-measures designs is the employment of multivariate analysis instead of the conventional univariate procedures (Myers, 1979). This approach has the advantage of circumventing certain important and frequently violated assumptions of the univariate approach. We highlight substitution of MANOVA for ANOVA procedure at the decision level.

Chapter 7 presents the investigator with a checklist for conducting MANOVA research. Essential steps and concepts for the effective use of MANOVA are extracted from different parts of the text and are arranged in sequential order so that the investigator can more readily relate MANOVA principles to the design of research, to research in progress, and to evaluating research already conducted.

A MANOVA problem completely hand calculated is provided in the appendix. This example may be skipped by the majority of readers without detracting from the decision features of the text. Those readers who feel comfortable with multivariate mathematics and are curious as to the precise steps underlying the various parts of the MANOVA solution may wish to consult the example at various points in the book.

1

Historical Origins of MANOVA

Era of Multivariate Techniques

The 1970s ushered in the multivariate era. Two major influences were responsible—the computer and the inherent inadequacy of the univariate approach. Although the necessary mathematics has been known for some time, the electronic computer with its fantastic speed first made multivariate techniques available for practical application. The influence of the computer has been supplemented by a growing recognition of the inadequacy of the strictly univariate approach to research. As one example of this inadequacy, Mowrer (1960) reports photographic recordings by Zener of a replication of Pavlov's classical conditioning study clearly showing that while the conditioned dog salivates to the sound of the bell as expected, the animal also "looks interested, hopeful, even 'happy,' and if not physically restrained will move bodily toward the place where the food is likely to be delivered" (p. 8). In keeping with the classic univariate approach, Pavlov presumably ignored the numerous concomitants of salivation as being irrelevant.

A further illustration of the inadequacy of the univariate approach is provided by current research on the effects of vitamin C on incidence of the common cold. Attention was initially centered on the single dependent variable—number of colds—but because of possible harmful side effects has subsequently shifted to another dependent variable, the effect of large doses of vitamin C on the kidneys.

Multivariate analysis techniques include the conventional multiple regression, the discriminant function and correlation, and the topic central to this text—multivariate analysis of variance. These techniques share the same basic conceptual structure, and for those investigators whose statistical training includes basic correlation and exposure to multiple regression, many important generalizations are possible and desirable.

A number of sources allude to the approaching multivariate era. In the preface to their text on multiple regression, Kerlinger and Pedhazur (1973) remark: "Within the decade we will probably see the virtual demise of one-variable thinking and the use of analysis of variance with data unsuited to the method. Instead, multivariate methods will be well-accepted tools in the behavioral scientist's and educator's armamentarium" (p. vi). In the preface to his book on behavioral statistics, Kerlinger (1973) makes even stronger assertions:

> The very nature of behavioral research is multivariate: Many variables act in complex ways to influence other variables. While some of the complexity can be handled with analysis of variance, it is only in the multivariate methods that the complexity of many psychological, sociological, and educational problems can be adequately attacked. We are in the midst of a revolution in research thinking and practice. Behavioral research is right now changing from a predominantly univariate emphasis to a multivariate emphasis. The change is extensive and profound. Even the nature of theory and problems is changing. . . .
> Multivariate analysis can be ignored or treated superficially only at the cost of the book being obsolescent, even obsolete. [Pp. x–xi]

Cattell (1952) very early in the present century asserted that behavioral research was complex and dynamic, demanding more complex research designs than were then extant. Cattell formulated new types of research designs that have been virtually ignored until recently. While Cattell was a visiting lecturer at Emory University in 1958 he was asked why the designs had been ignored by scientists. Cattell replied that no one understood them.

As an indication of the rapidly increasing popularity of multivariate techniques (including MANOVA), scarcely any issue of a scientific journal appears that does not use them in at least one report. Perhaps the most convincing indicator of their recognition is the appearance of a chapter on multivariate analysis in an elementary statistics textbook (Hinkle, Wiersma, & Jurs, 1979).

Sequential Trends in Application of Multivariate Techniques

By noting the sequential trends in older research methods, it is possible to gain an accurate, intuitive grasp of the nature of MANOVA. The reader may find it extremely helpful to study Figure 1–1 very carefully prior to and during the following discussion.

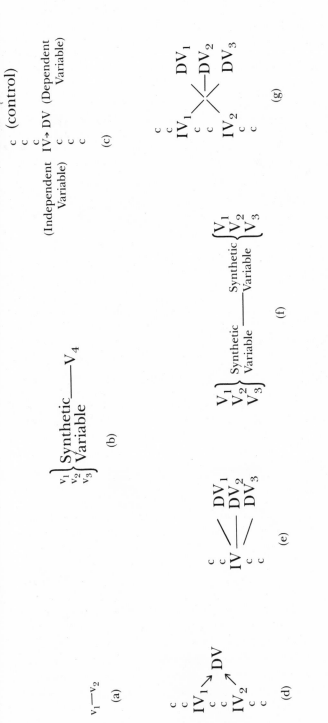

Figure 1–1. Schematized Views of Research Methods. They are: (a) Pearson r, (b) R, (c) experiment, (d) ANOVA, (e) discriminant function, (f) canonical correlation, (g) multivariate analysis of variance.

Pearson r

The Pearson product–moment correlation coefficient, one of the early statistical techniques ($r = \Sigma\, Z_x\, Z_y/n$), indicates the degree of relationship between two variables. Depending on the degree of relationship, the researcher may predict either variable on the basis of knowledge of the value of the other. However, the two variables have equal status in that either may be the cause of the other or both may be caused by some other variable(s). Recognition that causation cannot be inferred from simple correlations led experimental scientists virtually to abandon correlation methods and to seek out another class of statistical techniques. Figure 1–1a depicts the correlation relationship. Both variables are denoted by the letter *V* to emphasize their equal status.

Multiple Correlation (R)

Basically, *R* is a Pearson product–moment *r* indicating the degree of relationship between two variables. However, one variable is a newly created composite (synthetic) score resulting from a sum of weighted portions of each variable in the predictor set. The weights are so derived that the correlation between the synthetic variable and another (criterion) variable is a maximum. Again, the synthetic and the criterion variable are alike in that neither variable may be regarded as the cause of the other and both may result from some other causative influence. Figure 1–1b depicts this relationship.

Synthetic Variables

The concept of a composite (synthetic) variable is fundamental to all multivariate statistics, and therefore further explanation will be given. In multivariate statistics, conventional usage features numerous terms such as composite scores, factor scores, discriminant scores, canonical scores, orthogonal scores, and so forth to designate a kind of composite score that is associated with different multivariate procedures.

The authors favor the single-term synthetic variable to designate the numerous other terms used. The word *synthetic* is defined (see Webster and Teall, 1972) to mean "made of parts or elements combined in a new form; composed artificially; produced artificially." The definition refers to a "new" variable—not to an arbitrary hodgepodge or a combination of preexisting variables. This definition seems to describe adequately the composite scores generated in multivariate work. The synthetic scores of multivariate techniques are derived according to mathematical rules so that they relate consistently to other multivariate statistics.

A simple example will illustrate. Suppose an investigator has available two sets of measures that are to be used in predicting a third (criterion) variable. The data are contained in Table 1–1. Variables X_1 and X_2 are to be used to predict X_3. It is apparent that neither variable X_1 nor variable X_2 separately can predict the criterion perfectly. However, it may be possible to form a new synthetic variable from the two predictor variables such that the new synthetic variable predicts the criterion variable more accurately than either separate predictor variable. Consider adding one-fourth of the first predictor variable, X_1, to one-half of the second predictor variable, X_2. This step illustrates the use of weights applied to the predictor variables to form a new synthetic variable. A comparison of the new synthetic variable with the criterion variable indicates that the two variables are now identical. Therefore, the newly found synthetic variable can be used to predict the criterion variable perfectly.

Table 1–1. Development of a Synthetic Variable to Increase Accuracy of Prediction

Predictor variable		Synthetic variable	Criterion variable
X_1	X_2		X_3
4	8	5	5
8	2	3	3
4	6	4	4
16	4	6	6

The reader may wish to verify that the weights applied to the predictor variables were not arbitrary. If the data of Table 1–1 are submitted to a multiple correlation analysis, the obtained R will be 1.00, verifying that the criterion variable is perfectly related to the new synthetic variable. Furthermore, the weights that were used to determine the portion of each predictor variable to be combined into the synthetic variable will be precisely the B weights that are used in a regression equation for predicting the criterion variable. The B weight for the first variable will be .25, and the B weight for the second variable will be .50.

Introduction of Differential Status to Variables

An experimental arrangement imparts differential status to variables. Three major classes of variables are recognized: controlled, independent (or causative), and dependent (or outcome). The experiment is governed by the Law of the Independent Variable, which describes the relationship between variables as follows. In an experiment, all variables are held constant (under control) except one. This is the independent variable, and it is manipulated by the experimenter. Concomitant observations are taken on a resultant or dependent variable. Figuratively, the experiment represents a kind of microscope that permits the scientist to magnify the relationship between two variables in order to determine whether one variable (the independent) "causes" the other (dependent variable). Figure 1–1c depicts the classic experimental setup. Differential status is accorded to the variables, and the investigator may draw inferences regarding causation. Historically, one of the first statistical techniques to be utilized for the experiment is the classic t test for difference between means.

Analysis of Variance (ANOVA)

Fisher (1953) developed a statistical hypothesis-testing procedure known as ANOVA. It retains differential status of variables (control, independent, and dependent); however, paradoxically, it enables an investigator to employ more than one independent variable in one and the same experiment and still satisfy the requirements of the Law of the Independent Variable. That is, an experiment may be performed in which there are any number of independent variables, but the method allows for only one dependent variable. The causal relationship between the dependent variable and the separate independent variables may be assessed as well as the possible interaction of independent variables on the dependent variable. Figure 1–1d depicts this development as involving differential status of variables with two or more independent variables contained within the same experiment.

ANOVA via R

Ward and Jennings (1973), Cohen (1968), Overall and Spiegel (1969), and others introduced behavioral scientists to multiple regression as an alternate method of performing analysis of variance after first coding the independent variables. The procedure highlights the complete neutrality of statistical techniques as to the nature of the data being analyzed. That is, analysis may be performed whether or not one has performed an experiment. The critical elements underlying the experiment are the differential status of variables with particular emphasis on control of variables.

Discriminant Function

Fisher (1936) developed the discriminant function as a means of classifying individuals (plants, animals, people) into groups based on two or more independent measures taken on each individual. The measures are each weighted and are then summed to create a new synthetic variable that maximally separates the groups. McNemar (1955) observes that if the group classification is a dichotomy (e.g., sex, delinquent–nondelinquent, etc.), the discriminant function applied to group classification and the multiple regression techniques applied to the prediction of the same dichotomous variable (in coded form such as 0, 1, or 1, 2) produce equivalent results. (It should be noted that Hotelling's T^2 statistic is also equivalent to the significance tests of the discriminant function and the R under these circumstances.) This correspondence between multiple regression and the discriminant function reveals two aspects of the discriminant function: a prediction aspect and a test of statistical significance. Multiple regression results in a correlation coefficient that indicates prediction accuracy; an analogous statistic such as Wilks's lambda is typically used with the discriminant function. The exact correspondence between R and Wilks's lambda is evident in the following equation: $R^2 = 1 - \Lambda$. The standard F test is used with multiple regression to test the statistical significance of R. In similar fashion a chi-square test or F test has been associated with the discriminant function for the same purpose. The foregoing remarks apply precisely when the grouping variable is dichotomous because a restriction of one dimension attends the two points in space. With more than two groups, the groups may be separated along two or more dimensions, thus requiring more than one discriminant function.

Fisher's discriminant function is of interest in another surprisingly important way. The creation of a synthetic variable and the test of statistical significance are mathematically equivalent for the discriminant function and the simplest case of MANOVA. If one independent variable (no restriction is imposed on the number of levels) is employed in a research design and multiple dependent measures are taken, the discriminant function may be used to analyze the results both for statistical significance and for the effectiveness of group membership prediction. Figure 1–1e represents this situation. The design may consist of an experiment (or study) and may be analyzed either via multiple regression (if only two groups) or by the discriminant function. The general questions to be answered in such research are: Taking into account the intercorrelations between measures, do the profiles of variables separate the treatment groups significantly? If so, how effectively are they separated? This latter question is equivalent to asking how accurately one can predict treatment-group membership if one knows the profile of measures for each individual.

Canonical Correlation

The canonical correlation procedure may be viewed as a Pearson r between two synthetic variables. Weights are obtained for each variable in each of two (or more) different sets of variables such that the resulting new synthetic variables correlate maximally. A distinctive difference between the canonical model and those preceding it is that for the first time there is provision for more than one variable on both sides of the equation (see Figure 1–1f). Application of the technique might involve correlating a set of variables thought to represent home conditions as one set with performance on the subtests of the Wechsler Bellevue as another set. The technique takes into account relationships between the sets on one or more dimensions.

The canonical model represents our most encompassing statistical model. If, as Cohen (1968) has phrased it, multiple regression is a general data-analytic system, then canonical correlation may be viewed as the most general of all data-analytic systems. As noted by Knapp (1978), canonical analysis subsumes R as a simpler case in addition to subsuming the techniques of Pearson r, the t test, ANOVA, and the discriminant function. Knapp (1978) neglects to mention that canonical analysis also subsumes multivariate analysis of variance. The investigator who has available a good computer program for canonical correlation and knows how to code qualitative variables into the system can readily derive any of the subsumed statistics with the one program.

It is enlightening to consider how canonical correlation subsumes other statistics. Basically, the feat is accomplished by differential weighting of variables. Consider the following situation, in which a number of presumably causative variables are included in one set of measures and a number of outcome, or dependent, variables are available in a second set of measures.

1. *Pearson r.* A weight of 1.0 is accorded to one variable in one set and a weight of 1.0 is given to another variable in the second set. All other variables in the two sets are weighted 0.0 or are omitted. The resulting canonical correlation is a simple Pearson r.

2. *t test.* The independent variable is coded to represent treatment-group membership in the first set, and a weight of 1.0 is given to only one (dependent) variable in the second set. Any remaining variables in the two sets are weighted 0.0 or are omitted.

3. Multiple correlation. A synthetic variable is created from one set by differentially weighting the predictor variables so that they correlate maximally with a single variable weighted 1.0 in the second set. Other variables are weighted 0.0 or are omitted in the second set.

4. Discriminant function. Group membership is coded in one set of variables, whereas differential weighting of measured variables is accomplished to produce new synthetic variable(s) in the second set. Any remaining variables in the first set are weighted 0.0 or are omitted.

5. ANOVA. Any type of ANOVA design may be accomplished by coding independent variables in the first set and weighting the dependent variable of interest as 1.0 in the second set. All other variables in the second set are weighted 0.0 or are omitted.

6. MANOVA. Coded independent variables (or observational qualitative variables) in the first set are correlated with a synthetic variable created by differential weighting of variables in the second set (see Figure 1–1g).

Finally we should note that by using appropriate models (Overall & Spiegel, 1969), data involving both equal and unequal group n can be readily accommodated by the canonical program for any of the statistical techniques. Multivariate analysis of variance, then, essentially involves one or more independent variables (or qualitative observed variables) in the one set and two or more measured dependent variables in the second set (see Figure 1–1g).

2

Conceptual Theory
Underlying MANOVA

A very large amount of current scientific research suffers from a major flaw. Multiple dependent-variable measures are taken in the typical experiment or study and then a separate univariate ANOVA is performed on each separate dependent variable without taking into account the interrelationships between the dependent measures and the possibly increased alpha (Type I) error levels. Even though sophisticated analysis of variance designs are employed as a guide in the design and conduct of the research and in the analysis of data, multiple dependent variables are analyzed sequentially or separately in univariate fashion—each separate analysis failing to take the others into account. This state of affairs is highly analogous to the use of the standard t test to make unplanned comparisons between numerous nonothogonal pairs of group means. The problem arising from both cited examples is that alpha (Type I) error is inflated by such practices. The probability is increased (above the alpha level employed in the research) that the investigator will conclude that the treatments are effective in situations where pure chance is operating.

Several correct procedures to follow for a standard analysis of variance design involving multiple dependent variables have long been known. The standard solution calls for a multivariate analysis of variance, but the complexity and tediousness of the calculations have deterred all but the hardiest of souls. Now the electronic computer, guided by long-awaited programs, has removed this source of difficulty and has made it possible for the investigator not only to assess accurately the probability levels but also to view the emergence of new synthetic variables that may open up unexpected and exciting vistas.

One word of caution seems appropriate at this time. It concerns the temporal relation between the dependent variables. Before MANOVA can be appropriate, the dependent measures must be conceptualized as being taken simultaneously. If systematic time lapses occur between measures, "causal" relations may take place among the alleged dependent variables, thereby confusing the status of variables. If such a data set is under examination, the use of causal modeling may be a viable alternative method of analysis. Kenny (1979) presents an excellent in-depth treatment of this topic.

Parallels between Univariate ANOVA and Multivariate MANOVA

There are striking parallels between ANOVA and MANOVA. We have noted previously that MANOVA is essentially ANOVA but with multiple dependent variables: that is, all measures are administered to each subject. The presence of more than one dependent variable introduces the possibility of varying degrees of correlation between the dependent variables. Assuming that the intercorrelations between all dependent variables are zero (at both the pooled error source and the treatment levels), the MANOVA approach simply sums the F ratios that would result from individual ANOVAs applied to the separate dependent variables (Li, 1964). Indeed, the common practice of repeating ANOVAs for each separate dependent variable would then be an appropriate analysis procedure, because the significance tests are orthogonal. (See chapter 3 for a discussion of statistical significance tests.) It would, of course, be extremely rare for this situation to occur with real data. Some degree of intercorrelation between dependent variables is virtually a certainty. Therefore, the appropriate approach must take into account that dependent variables to some degree measure the same thing. Furthermore, it is possible that a variable that by itself is useless in measuring treatment effects will be valuable in removing error variance from another variable that separates treatment groups, thus permitting the effective variable to make an even more effective contribution. This type of variable is the well-known classic suppressor variable encountered in multiple regression. A more detailed examination (at the conceptual level) of ANOVA and MANOVA will highlight similarities between the two procedures.

1. *Univariate ANOVA*. In the classic analytic tradition, the extent of variability in the data set is first determined. Assuming a one-way ANOVA, this total is analyzed into two additive portions: a treatment component and an error component.

$$SS_{total} = SS_{treatment} + SS_{error}$$

where SS is used to designate sum of squares.

Two important ratios are formed from components of the sum of squares. The first is the classic F ratio, which pits the treatment portion against the error portion.

$$F = \frac{SS_{treatment}/df_{treatment}}{SS_{error}/df_{error}}$$

This ratio enables the investigator to utilize probability expectations in order to assess the reliability of the results.

The second ratio pits the treatment sum of squares against the total sum of squares.

$$\text{Correlation ratio } (\eta^2) = \frac{SS_{treatment}}{SS_{total}}$$

This simple, intuitive ratio describing the proportion of the total variation produced by the treatments allows the investigator to assess the relative strength of the relationship (substantive significance) between the research treatments and the dependent variables.

2. *Multivariate MANOVA*. The use of two or more dependent variables in an ANOVA framework requires that the cross-products between dependent variables as well as the conventional sums of squares for each separate variable be taken into account. It will be helpful to use the symbols $SSCP$ to represent the sum-of-squares-and-cross-products matrix. A direct correspondence with ANOVA is apparent in that a total $SSCP$ matrix is computed for the dependent variables and this matrix of totals is decomposed (analyzed) into two major components: a $SSCP$ matrix for treatment effects and a $SSCP$ matrix for error. The analysis can be meaningfully represented as for ANOVA.

$$SSCP_{total} = SSCP_{treatment} + SSCP_{error}$$

It is helpful to examine this relationship in more detail.

$$
\begin{array}{ccc}
\underline{SSCP_{total}} & \underline{SSCP_{treatment}} & \underline{SSCP_{error}} \\
\begin{array}{c c c} & V_1 & V_2 \\ V_1 & - & \\ V_2 & & - \end{array}
&
= \begin{array}{c c c} & V_1 & V_2 \\ V_1 & - & \\ V_2 & & - \end{array}
&
+ \begin{array}{c c c} & V_1 & V_2 \\ V_1 & - & \\ V_2 & & - \end{array}
\end{array}
$$

The matrix of $SSCP_{total}$ is shown decomposed into two matrices: $SSCP_{treatment}$ and $SSCP_{error}$. The diagonal entries contain the sums of squares that are traditionally formed into univariate F ratios. The off-diagonal entries represent the sums of cross-products components, which take into account the degree of correlation between variables. In essence, as one moves from ANOVA to MANOVA, one moves from single elements that are additive to matrices consisting of elements that are additive.

Just as in ANOVA, in MANOVA two important ratios may be formed: a statistical test of significance, which may take one of several forms, and a test of substantive significance, similar to eta-square. Consider first the test of statistical significance. In a fashion analogous to ANOVA, the $SSCP_{treatment}$ is "divided" by the $SSCP_{error}$. To accomplish division with matrices, it is first necessary to invert the matrix ($SSCP_{error}$) to be used as a divisor (matrix inversion is the counterpart of transforming a number to its reciprocal in scalar math). The inverted matrix is then postmultiplied (matrix multiplication) by the $SSCP_{treatment}$ ($SSCP_{error}^{-1} \cdot SSCP_{treatment}$). A new matrix results that reflects the extent to which the variables discriminate between the treatment groups. At this point, an important issue arises with respect to the number and nature of different dimensions that underlie the dependent variables separating the treatment groups. A method with the generic name *factor analysis* is used to answer these questions. In order to acquaint the reader with important links between this method and MANOVA, we shall briefly discuss several basic concepts of factor analysis.

Factor Analysis and MANOVA

It is ironic that a method of factor analysis (principal components) lies at the very heart of MANOVA. An investigator using a computer statistical program for MANOVA could easily be unaware that the mathematical operations of factor analysis are performed by the program. MANOVA represents the ultimate in experimental analysis while incorporating correlational procedures. As Cronbach (1957) notes, the experimentalist sought to abandon correlation techniques and to use only the classic tests of statistical significance such as ANOVA. The primary role played by factor analysis in MANOVA is that of determining the dimensionality and nature of the sets of dependent variables that are sensitive to treatment effects, thereby making it possible to apply the ANOVA techniques and interpretations along separate, independent dimensions.

In order to clarify further the role of factor analysis in MANOVA, we shall briefly examine the nature of three outcomes of conventional factor analysis—factor loadings (correlations between variables and a factor), eigenvalues (to be explained later), and synthetic scores—at a conceptual level. We shall also consider the corresponding outcomes of the application of factor analysis to MANOVA data. Table 2–1 provides factor-

Table 2–1. Hypothetical Factor Analysis of Intercorrelation Matrix

Variable	Factor I	Factor II	h^2
1	.7	.1	.50
2	.5	.1	.26
3	.1	.6	.37
4	.6	.2	.40
5	.2	.5	.29
6	.1	.6	.37
Eigenvalue (λ)	1.16	1.03	2.19

analytic results for a hypothetical example. It is assumed that a correlation matrix between six variables has been factor analyzed into six principal components. In order to simplify the discussion, we have shown in the table only the first two principal components (factors). Factor loadings appear as elements in the table, e.g., Variable 1 has a factor loading of .7 on the first factor and a loading of .1 on the second factor. These factor loadings represent Pearson product–moment r's between the scores on the variable and each of two factor scores.

The two factor scores (synthetic variables) result from using a form of the factor loadings as weights in combining the six variables into two new scores. Table 2–2 contains the normalized weights, which are multiplied by the original variables and are summed in order to obtain factor scores. The normalized weights are computed by dividing each factor loading by the square root of its associated eigenvalue. The new factor scores are uncorrelated with each other and therefore represent independent dimensions.

The new synthetic variables (factor scores) are interpreted by noting the common characteristics of variables that correlate (and the direction of the correlations) with the factors. In the hypothetical example, the variables that correlate principally with the first factor (Variables 1, 2, and 4) may be quantitative in nature, whereas those variables loading highest on the second factor (Variables 3, 5, and 6) may be clearly of a verbal nature.

The eigenvalue, to which we alluded in deriving normalized weights for determining factor scores, is a very important concept for both factor analysis and MANOVA. Table 2–1 displays the two eigenvalues, corresponding to the two factors, at the base of the factor columns. An eigenvalue indicates the portion of variance of the variables that is in common

Table 2–2. Normalized Weights Used to Compute Synthetic (Factor) Scores

	Factor	
Variable	I	II
1	.650	.099
2	.464	.099
3	.093	.591
4	.557	.197
5	.186	.493
6	.093	.591

Note: Factor weight = Factor loading / $\sqrt{\lambda}$.

with the factor. For example, squaring a factor loading of a variable produces r^2, which indicates the proportion of variance of the variable predictable by or in common with that factor score. If each variable's loading on a factor is squared and summed, the eigenvalue results.

Thus the eigenvalue is essentially a total of the portions of variance of all the variables that can be accounted for by the factor score. The eigenvalue for the first factor in Table 2–1 is 1.16, indicating the total portion of variance of the six variables that is in common with the factor score. In principal-components factor analysis, the variances of the variables are expressed in standard form ($S^2 = 1.0$), and ones are entered in the diagonals of the correlation matrix, which is then factored. In the example, the total variance (trace) of the variables is 6.0, and the first factor accounts for approximately 19% of the (trace) total variance of the variables. It is obvious that Factor II is a weaker synthetic variable, able to account for approximately 17% of the trace.

In contrast to the intercorrelation matrix, which is the target in conventional factor analysis, in MANOVA factor analysis is applied to a matrix that originated from sums of squares and cross-products. Readers who find this procedure strange will find Nunnally's discussion (1962) of the application of factor analysis to matrices other than correlations particularly enlightening. As we indicated earlier, the matrix to be factored in MANOVA represents the extent to which variables continue to covary and separate the treatment groups after "error" variability of individual variables and covariation between variables is taken into account. The nature of the factor solution is such that the magnitude of separation between treatment groups diminishes with each succeeding factor extracted. A chi-square test of statistical significance of group separation on each synthetic variable is available.

The synthetic variables in MANOVA resulting from factor analysis may be regarded as independent, uncorrelated dimensions along which treatment-group means (centroids) are maximally separated. The reader can verify the uncorrelated nature of the synthetic variables by computing Pearson r's between them, using pooled cross-product matrices from the error sources. Assuming three or more treatment groups, Pearson r between treatment-group centroids on pairs of synthetic variables will also be found to be zero.

The factoring procedure in MANOVA is applied to a product matrix $(SSCP_{error}^{-1} \cdot SSCP_{treatment})$. The factor loadings that result are not correlations, as in conventional factor analysis, but are instead relative weights of the original dependent variables in forming the new synthetic variables. Table 2–3 contains hypothetical relative weights and associated eigenvalues of the dependent variables from factor analysis of the matrix $(SSCP_{error}^{-1} \cdot SSCP_{treatment})$. Note that (as in conventional factor analysis) the eigenvalues represent the sum of squared values of the associated weights. The weights are normalized, as in conventional factor analysis, and are shown in Table 2–4. These normalized weights are used to derive uncorrelated synthetic scores.

Table 2–3. Weights Resulting from Application of Factor Analysis to Data Matrix
$$(SSCP^{-1}_{error} \cdot SSCP_{treatment})$$

	Factor	
Variable	I	II
1	3.0158	−.2108
2	−1.3453	.6714
Eigenvalue (λ)	10.9049	.4952

Table 2–4. Normalized Weights for Computing Synthetic Scores

	Factor	
Variable	I	II
1	.913	−.300
2	−.407	.954

Note: Discriminant weight = Factor loading / $\sqrt{\lambda}$.

If the relative weight for each variable is multiplied by the standard deviation of the corresponding variable, the resulting product may be used to indicate the relative importance of the variable in forming the new synthetic variable. This procedure may prove useful in interpreting the nature of the new synthetic variables.

Another approach is available for interpreting the nature of those synthetic variables that separate the treatment groups significantly. The approach involves consideration of the correlation between a particular synthetic variable and each of the dependent variables. In this case the matrix resembles the factor loading matrix of conventional factor analysis of variables. However, it is important that the correlations be determined from pooled cross-product matrices from within the error sources of the research design. In this way, the resulting correlations are not inflated by different mean levels of the dependent variables that are produced by the treatment effects. The logic of this procedure is analogous to that used in analysis of covariance, where the correlation used to "correct" the dependent variable for its relation to the covariate is derived from pooled cross-products from within the error sources. Similarly, in a one-way MANOVA design, the $SSCP$ matrix between dependent and synthetic variables is computed for each error source separately and is then pooled for all error sources. Therefore, between-groups treatment effects for the dependent variables do not appear in the final correlation between each dependent and synthetic variable.

If the pooled $SSCP_{error}$ matrix for the dependent variables were converted to a correlation matrix, it would be appropriate to apply factor analysis to explore the factor structure of the variables unaffected by the treatments. Thus, assuming homogeneity of the error dispersion matrices, hypotheses regarding the number and nature of dimensions underlying the dependent variables can be evaluated.

The nature of the synthetic variables may be shaped by a variety of patterns of treatment effect; we shall discuss two interesting examples. First, suppose the dependent variables consist of several measures of each of two uncorrelated dimensions. A factor analysis of the pooled error sources will reveal the two factor patterns of loads of the variables. Further suppose that the treatments affect the two measured dimensions very differently. Then the intercorrelations between the dependent variables (calculated from pooled error sources) and the synthetic variables will faithfully reflect the two quite different dimensions of the dependent variables. In this situation, the two synthetic variables are likely to separate the treatment-group centroids in a statistically significant manner. The pattern of correlations between the dependent and synthetic variables will resemble the factor loadings resulting from the earlier factoring of the dependent variables.

Still assuming the two-dimensional example just cited, a second type of outcome might involve correlations of similar magnitude for all dependent variables on the first and succeeding synthetic variables. In this instance, the treatments may have influenced all measured dependent variables similarly. Research involving the classic Hawthorne effect might well produce a similar discriminant effect on all measured dependent variables. A factor analysis of the r matrix between dependent variables resulting from pooled error sources would reflect the dimensionality of the dependent variables. In this instance, the first synthetic variable would account for most of the treatment-group differences, and it is unlikely that successive synthetic variables would separate the treatment groups significantly.

This brief discussion of the application of factor analysis to conventional data in contrast to MANOVA data will assist the reader to examine the essential nature of the four classic MANOVA tests of statistical significance more meaningfully.

MANOVA Tests of Statistical Significance

1. *Wilks's lambda criterion* $[\Lambda = \Pi (1 + \lambda_i)^{-1}]$. The reciprocals of the quantity (1 plus each eigenvalue) are multiplied in succession. The numerical value of this statistic is routinely converted to an approximation of the conventional F ratio. This ratio is sometimes designated R in honor of Rao (Tatsuoka, 1971, p. 200).

$$R = \frac{1 - \Lambda^{1/s}}{\Lambda^{1/s}} \cdot \frac{ms - pdf_h/(2+1)}{pdf_h}$$

$$m = df_e + df_h - (p + df_h + 1)/2$$

$$s = \sqrt{\frac{p\,(df_h)^2 - 4}{p^2 + df_h^2 - 5}}$$

where p = number of dependent variables, df_h = degrees of freedom for treatment tested, and df_e = degrees of freedom for error.

2. *Lawley–Hotelling trace criterion* $(\tau = \Sigma\lambda_i)$. The eigenvalues obtained from factoring the "discriminating" matrix $(SSCP_{error}^{-1} \cdot SSCP_{treatment})$ are simply summed and are referred to special tables (Pillai, 1960) for interpretation of statistical significance.

3. *Roy's greatest characteristic root criterion, or GCR* $[\theta = \lambda_1/(1.0 + \lambda_1)]$. This test is unlike the others in that a decision as to the reliability of the results is based on only one (the first) dimension (eigenvalue) of the measurements. The other three criteria employ all of the obtained dimensions in the test of statistical significance.

4. *Pillai's trace criterion* $(V = \Sigma \theta_i)$. Roy's formula is applied to each of the eigenvalues, and the results are summed across the roots.

The appropriate tables are then consulted to determine significance. For the Lawley–Hotelling trace criterion, Roy's greatest characteristic root, and Pillai's trace criterion, the tables are entered according to three parameters S, M, and N.

$$S = \min (df_h, p)$$
$$M = (|df_h - p| - 1)/2$$
$$N = (df_e - p - 1)/2$$

where df_h = degrees of freedom associated with the treatment effect, df_e = degrees of freedom associated with the error term, and p = number of dependent variables. The test of substantive significance (analogous to eta-square) is most easily illustrated by examining an alternate way of computing Wilks's lambda. The formula for Wilks's lambda may be expressed in another manner as $\Lambda = |SSCP_{error}| / |SSCP_{total}|$, where vertical lines on each side of $SSCP_{error}$ and $SSCP_{total}$ denote determinants. For the moment, note that the ratio being computed is intuitively opposite to that of eta-square. Eta-square represents the proportion of total variance accounted for by the treatment effects. But in Wilks's lambda, the ratio indicates the proportion of total $SSCP$ matrix (determinant) *not* accounted for by the treatment effect. By subtracting Wilks's lambda statistic from 1.0, we obtain the statistic analogous to eta-square.

It is also instructive to consider the numerator and denominator of Wilks's lambda statistic. The vertical bars bordering the symbol $SSCP$ indicate that a single value (scalar) is derived to represent the particular matrix. This single value is labeled a determinant. It is computed by multiplying in succession the eigenvalues extracted from the matrix and bears a remarkable resemblance to more conventional procedures for determining areas or volumes of geometric figures. For example, consider the area of a rectangle that is determined by multiplying length by width. A rectangle contains two dimensions (factors)—length and width—and their product is the area. Likewise, consider how the volume of a rectangular solid is determined. In this instance, there are three dimensions (factors) required—length, width, and height. The volume is determined by multiplying length by width by height. From these illustrations it is apparent that the determinant of a matrix of dependent variables represents the amount of shared variance or information contained in the dimensional structure.

Differential Sensitivity of Test Criteria Related to Distribution of Trace

An important question arises concerning the sensitivity of the several MANOVA test criteria as a function of how the trace is distributed into the eigenvalues. In MANOVA, *trace* refers to the discriminating variance of the variables undergoing factor analysis. The resulting eigenvalues (if all are extracted) will sum to the trace. The trace could, for example, be almost totally located in the first eigenvalue. This feature would indicate only one strong factor (dimension). At the opposite extreme, the trace could be rather equally distributed among the eigenvalues. Intermediate distribution of trace would likewise be possible. The basic question then arises as to what effect, if any, differential distribution of trace into the eigenvalues will have on the various tests of statistical significance.

Although complex computer simulations have been used to study the sensitivity of test criteria as related to distribution of trace (Schatzoff, 1966; Olson, 1974, 1976), we performed one such study that required only extant multivariate tables (Barker & Barker, 1979). The results of all studies indicate quite clearly that the four different test criteria are differentially affected by distribution of trace. As expected, concentration of trace into the first root causes Roy's greatest characteristic root to be the most sensitive test, whereas equal distribution of trace into all roots makes Wilks's lambda the most sensitive. Lawley–Hotelling trace and Pillai's trace criteria are intermediate in sensitivity in both instances. Barker and Barker (1979) interpreted their results as indicating the need to take differential test sensitivity into account in the design stage of research by selecting a priori the appropriate test statistic. However, because of the complexity of the several issues involved and possible value to the reader, it seems desirable to report all four tests in any MANOVA study. It is clear from the study that investigators need to formulate some research hypotheses regarding the dimensionality and nature of the dependent variables under study.

Assumptions Underlying ANOVA and MANOVA

The two basic statistical assumptions of ANOVA are: normality of distribution in the population of the dependent variable and homogeneity of variance of the sources from which the error term is derived. A classic study by Norton (1952) systematically violated these two assumptions in order to note the effect on the F ratio. The essential findings were that departures from normality of distribution (even gross ones) had little effect on the F distribution. Similarly, mild to gross departures from homogeneity of variance had surprisingly little effect on the F distribution. These findings led eventually to the virtual abandonment by researchers of tests of homogeneity of variance in published research. ANOVA is termed robust with respect to violations of assumptions underlying it.

Before we leave this issue, we should note that Norton's study did not include two very common occurrences in research: unequal number of observations in the error sources and heterogeneity of variance of the error sources due to an outlier error component. The effect of unequal cell sizes is very complicated and has been widely documented as being reason for particular concern. Therefore it will not be discussed further. In contrast, little attention has been given to heterogeneity of variance produced by an outlier error component. As we discuss this problem we will assume that the error sources contain an equal number of observations.

We should note that the mean square for error in ANOVA is a *mean* (average) and that an average is strongly affected by an outlier measurement. Furthermore, the direction of effect exerted depends on the direction of the outlier. In other words, lack of homogeneity of error variance in an ANOVA design involving an outlier may lead to an error mean square that either is inflated or is too small, depending on the direction of the outlier. Under these circumstances the results of the F test are either too conservative or too sensitive.

It seems that the careful investigator should test routinely for homogeneity of variance. If lack of homogeneity is found, a simple plot designed to spot an outlier (or outliers) should follow. If no outlier is found and the distribution of error sources is fairly symmetrical, one may take comfort in the Norton study. However, the discovery of a major outlier should at least be recognized in the interpretation of the F ratio (whether it is likely to be too conservative or too sensitive). Other alternatives available to the investigator include omitting the offending treatment group (as in a one-way ANOVA), transforming the dependent variable measures to a new scale that produces homogeneity of variance, or perhaps shifting to a less powerful, nonparametric statistical technique. Also, one should not neglect to consider the possibility that the treatment effect under scrutiny may affect variability rather than the mean, a finding that may be of considerable interest.

The assumptions of MANOVA closely parallel those of ANOVA. MAN-

OVA assumes multivariate normality of distribution and homogeneity of dispersion matrices. The dispersion matrices refer to the error sources from which the $SSCP_{error}$ is pooled. For example, in one-way MANOVA, there are two or more measures on each individual in each treatment group. A $SSCP$ matrix is determined for each treatment group; then these matrices are summed to produce the $SSCP_{error}$ matrix. If there exists a gross outlier group such that one $SSCP$ matrix is extremely different from the others, it will exert an effect on the error term analogous to that discussed for ANOVA. Tests of homogeneity of dispersion matrices, for example, the Bartlett test (Timm, 1975) or the Box test (Box, 1949), convert the separate error sources to determinants and then compare the homogeneity of the determinants. If a significant heterogeneity of dispersion matrices is detected, it is appropriate to plot the determinants to ascertain if an outlier is present. If an outlier is clearly present, one may wish to use one of the several strategies mentioned for ANOVA above.

Aside from the caution regarding outlier matrices in MANOVA, the scientific evidence on the effect of lack of homogeneity of dispersion matrices appears very similar to that found by Norton in ANOVA. Ito and Schull (1964) and Ito (1969) conclude that MANOVA is robust with respect to lack of homogeneity of dispersion matrices and there should be little concern—especially when each treatment unit has an equal number of individuals.

There appears to be no compelling reason to be overly concerned about the assumption of normality of multivariate distribution. It has long been known that multivariate distributions such as those associated with the distribution of the synthetic variable in multiple regression are apt to be normally distributed even though the individual variables composing the synthetic variable are not. The same kind of relationship characterizes synthetic scores in factor analysis and the individual variables that constitute them.

To this point, the discussion about the effect on MANOVA of failure to meet assumptions has been quite general. If the MANOVA design structure permits the extraction of only one eigenvalue from the product matrix ($SSCP_{error}^{-1} \cdot SSCP_{treatment}$), all four MANOVA criteria are equivalent. If more than one eigenvalue is extracted, the MANOVA tests are not equivalent, and the failure to meet MANOVA assumptions may exert a different effect on the different tests. To complicate matters further, the extent to which the eigenvalues differ in magnitude may play a role. Relative differences in magnitude of the eigenvalues are generally denoted in two ways. If the trace (sum of eigenvalues) is mostly contained in the first eigenvalue, it is termed a *concentrated* structure. If the magnitude of the eigenvalues is relatively alike, it is termed a *diffuse* structure.

Olson (1976) studied the effects of nonnormality and heterogeneity of dispersion matrices on each of the four MANOVA test criteria. Nonnormality was found to exert little effect on any of the four tests. Heterogeneity of dispersion matrices exerted a different effect on the four MANOVA tests. Inflation of Type I error was particularly large for Roy's greatest characteristic root, and on the assumption that most behavioral research involves multidimensional effects (diffuse structure), Olson asserts that Roy's greatest characteristic root is not a desirable MANOVA test. The other MANOVA tests performed similarly for large samples. For small samples, of the three remaining tests, Pillai's trace was found to be least affected by heterogeneity of dispersion matrices while retaining desirable power characteristics.

Stevens (1979) agreed with Olson's conclusion regarding the robust nature and power characteristics of Pillai's trace under conditions of diffuse structure. However, Stevens argues that concentrated structures are prevalent in behavioral research. He insists that the degree of heterogeneity of dispersion matrices studied by Olson is far more severe than is typical. Under more usual degrees of heterogeneity, and assuming concentrated structure, Stevens asserts that the four MANOVA criteria are essentially equivalent in Type I error and that therefore the choice of a test criterion should be based on considerations of statistical power.

Harris (1975) favored Roy's greatest characteristic root over other MANOVA criteria. Harris defends this view for reasons unrelated to the present discussion. Harris notes that a significant Roy's greatest characteristic root coincides with significant differences between group centroids on the first synthetic variable. Statistical significance on the other three MANOVA tests may coincide with no statistically significant difference between group centroids on any single synthetic variable.

From the preceding discussion, it is apparent that the selection of an appropriate MANOVA test relates to complicated issues of robustness, statistical power, and certain consistency relationships inherent in the model.

3

Decision Strategies

Several different strategies are used to deal with the MANOVA situation in which multiple dependent variables appear in a design that otherwise resembles ANOVA. The chief strategies are the Bonferroni t, classic MANOVA analysis, the Hummel–Sligo procedure, and a mixed strategy. This chapter focuses on these four in detail, after first briefly reviewing statistical decision errors and the power concept, both of which apply to both ANOVA and MANOVA.

Decision Errors

Type I Error

A Type I (alpha) error represents a wrong decision that the research treatments under study produced the differential outcomes observed in the dependent variable(s), when actually the differences were produced by chance or random events. This type of error is readily controlled (but is never eliminated when statistics is used) by selecting for the test statistic a significance level that represents a rare chance occurrence. Traditionally this level has been either .05 or .01. Unless the test statistic attains or exceeds the critical level, the chance, or null, hypothesis is retained.

Control of Type I error (alpha) is straightforward when two treatment groups and one dependent variable are involved. Student's t was designed for such a situation, and tabled values for different significance levels are readily available. When more than two treatment means are compared, problems arise with respect to the actual alpha level that should govern each separate comparison. These brief remarks introduce the difficult area of multiple comparisons. Readers who wish to explore the area in greater detail can find extensive reviews by Ryan (1959), Kirk (1968), Myers (1979), and Keppel (1973).

The topic of multiple comparisons will be discussed only briefly in order to clarify some basic issues and to acquaint the reader with the authors' views on multiple comparisons, which are applied throughout the book. Controversy in the area of multiple comparisons centers on three major issues: (a) the nature of independence and orthogonality, (b) planned (a priori) versus unplanned (post hoc) comparisons, and (c) the appropriate alpha level to be used in making each separate comparison. Under certain conditions, a series of individual comparisons may each be performed at a chosen alpha level (e.g., .05). Under other circumstances, the alpha level may apply to an entire set of comparisons such that each individual comparison is made at an alpha level less than .05. An example of the latter technique is the Newman–Keuls approach (Kirk, 1968), which compares ordered treatment means against each other while holding the probability for the entire set of comparisons at the chosen alpha level. In order to accomplish this task, the individual comparisons are each made at a more demanding alpha level.

The decision as to whether to apply the chosen alpha level to each separate comparison or to an entire set of comparisons depends partly on whether the separate comparisons are independent of each other. Unfortunately there is some controversy as to what constitutes independence. Consider the following situation on which there is general agreement. Two different experimenters, each testing two different pairs of treatments, conduct their experiments at different points in time. The two experiments are viewed as clearly independent, and the alpha level is set at .05 for each comparison. This situation may be viewed as an example of *simple* independence.

In contrast to the previous example, investigators customarily study the effects of more than two treatments in the same experiment. In order to highlight the issue under consideration, dependence due to use of the same error term in performing multiple comparisons will be ignored in this discussion. Obviously the several possible comparisons between treatment means do not reflect simple independence. Under this type of circumstance, the standard alpha level may not be appropriate for each separate comparison. However, in a mathematical sense, for each degree of freedom available in ANOVA, there exists an orthogonal (independent) comparison. Mathematical independence implies that the outcome of an orthogonal comparison is not deducible from the set of orthogonal (independent) comparisons.

Problems arise from regarding mathematical orthogonality as equivalent to simple independence. Within a set of data there exists more than one set of orthogonal comparisons. The investigator is confronted with the need to select a particular orthogonal set. For this reason, some methodologists insist that the use of a standard alpha level (e.g., .05) per comparison is not appropriate unless the investigator selects the particular orthogonal set on a priori grounds. For unplanned orthogonal and nonorthogonal comparisons, the alpha level per comparison should take into account the entire set of comparisons being made. As a result it is necessary to alter the alpha level per comparison to a more demanding level. Still other methodologists recommend that the omnibus F test of treatment effect attain significance before any individual comparisons are deemed appropriate. Not all methodologists agree that mathematical orthogonality is equivalent to simple independence. Therefore other strategies have been developed for keeping alpha level under control as multiple comparisons are made in a set of data.

The previous discussion highlights a portion of the controversy in the area of multiple comparisons. The disagreements arise from theoretical and empirical considerations that may not soon be resolved. With respect to post hoc comparisons, Keppel (1973) laments: "The unfortunate aspect of all this is that no approach to the problem is logically correct" (p. 156). If alpha level is controlled for the entire set of comparisons, the separate individual comparisons must be conservative and must thus introduce compensatory Type II error. Kirk (1968) regards this issue as the crux of the matter in selecting a multiple-comparison strategy. The investigator should choose a multiple-comparison strategy such that the desired alpha protection level is maintained for the set of comparisons and at the same time provides minimal inflation of Type II error.

This cursory review of the multiple-comparison issue provides a background for informing the reader of the strategies endorsed by the authors. In view of the subjectivity inherent in the definition of a planned comparison and in the selection of an orthogonal set of contrasts, the authors endorse McNemar's (1969) approach, which requires a significant omnibus F before multiple comparisons are performed. If the omnibus F is significant, then planned, orthogonal comparisons may each be evaluated at the standard alpha level. For nonorthogonal and/or unplanned comparisons, Scheffé's t test is deemed appropriate for making any and all comparisons. Investigators who regard the Scheffé test as too conservative may double the standard alpha level for each comparison (Ferguson, 1981; Scheffé, 1959). The Scheffé t is robust under nonnormality, heterogeneity of error variance, and unequal n per treatment condition.

The Scheffé technique involves computing a traditional t ratio and then calculating a Scheffé value to replace the standard t table value required for significance. Although the Scheffé procedure adequately controls Type I error for any number of comparisons involving a single dependent variable, it is not viewed as being appropriate when additional dependent variables are used in the same research design (MANOVA). (An exception to this approach involves the use of orthogonal measures and application of the Scheffé method to each dependent variable separately, which appropriately takes into account unequal error estimates for the different dependent variables.)

Another important method of dealing with multiple comparisons was developed by Bonferroni (Dunn, 1961; Perlmutter & Myers, 1973). Interest is centered on a family or set of multiple comparisons. Individual comparisons need not be orthogonal, but the set of comparisons should be formulated a priori. Control of Type I error at the desired alpha level is accomplished by simply dividing the chosen alpha value by the number of multiple comparisons to be performed. By the additive law of probability, one is then assured that the probability that either the first or the second (etc.) of the comparisons will be significant at the chosen level is correct. The approach is entirely analogous to the practice of calculating the probability of rolling a 1 or a 2 on one throw of a fair die. In this example, probability is computed by adding the separate probabilities. The Bonferroni method is extremely versatile and will be considered later in the context of MANOVA data. Comparisons need not be limited to one dependent variable.

Type II Error

A Type II (beta) error refers to a situation in which the decision maker concludes that chance (random elements) produced the observed differences, when in actuality the treatment conditions produced the observed outcomes. Type II error is quantified as the probability of making the incorrect decision. Like the Type I error, this error can be specified and controlled by the investigator under conditions that we shall discuss in connection with statistical power.

Statistical Power

Statistical power refers to the probability that chance will be correctly rejected (the null hypothesis) when it is actually false. In other words, observed differences in treatment values are assumed to be related to treatment effects, and the investigator correctly draws that conclusion, thereby rejecting chance as an explanation. Note that statistical power may be quantified by subtracting the value of Type II error from 1.00.

The procedure for estimating the statistical power of an experiment has long been available (Tang, 1938; Pearson & Hartley, 1951) but is rarely used. However, the use of statistics as a tool for decision making demands that it be taken seriously (Cohen, 1969). Numerous good discussions with worked examples are available on this topic, for example, Senders (1958), Myers (1966), and Kirk (1968). The following discussion is intended to provide a common rationale for the estimation of statistical power for both ANOVA and MANOVA, but major emphasis is placed on power estimates for MANOVA.

ANOVA Power Analysis

In univariate ANOVA, under the null hypothesis the treatment mean square divided by the error mean square (F ratio) follows the tabled F distribution. If the null hypothesis is false, the F ratio follows a different distribution known as noncentral F, which may be used to estimate statistical power. Three parameters are required in order to make use of this noncentral F distribution: degrees of freedom for the treatment mean square (df_h), degrees of freedom for the error mean square (df_e) and delta.

$$\delta^2 = \frac{\sum_{j=1}^{k} n\ \beta_j^2}{\sigma_e^2}$$

where

$$k = \text{number of treatment groups}$$
$$n = \text{number of subjects per treatment unit}$$
$$\sum_{j=1}^{k} n\ \beta_j^2 = \text{sum of squares for treatment effect}$$
$$\sigma_e^2 = MS_e = \text{the mean square for error in the denominator of the } F \text{ ratio}$$
$$\beta_j = (M_j - M) = \text{a single difference between a treatment mean and the mean of all treatments}$$

Tables were developed by Tang (1938), and power curves were charted by Pearson and Hartley (1951) to simplify the estimation of statistical power. In order to use either the tables or charts a quantity (phi) must be calculated and entered into a table or chart corresponding to degrees of freedom associated with the F ratio.

$$\phi = \left(\frac{\delta^2}{df_h + 1} \right)^{1/2}$$

where df_h = degrees of freedom for the treatment effect

The phi value may be used for any type of ANOVA design. It is frequently used to make a post hoc estimate of statistical power in an experiment in which the null hypothesis is retained. The value of phi may be useful in the research planning stage. Using prior research findings or pilot research, estimates of treatment effects and error variance associated with a given sample size may be inserted in the equations to estimate the statistical power of an experiment yet to be performed. If the power estimate is not adequate, the investigator may consider one of two steps, either separately or jointly.

1. Values of n of increasing size may be inserted consecutively into the equation and phi may be recalculated until the power estimate is satisfactory.

2. More efficient research designs may be considered that will have the effect of decreasing the error variance (σ_e^2); for example, instead of a one-way design, one might consider using a Treatments × Subjects design, a Treatments × Levels design, or perhaps an analysis of covariance design. Prior research or pilot data, using alternative designs may provide a reduced error estimate (σ_e^2), which can be used for power estimation.

MANOVA Power Analysis

As noted earlier, for the two-treatment-group case, MANOVA data can be analyzed in either of two conventional ways: by discriminant analysis or by multiple regression. In either case a single new synthetic variable is derived in place of the original dependent variables. Since a single new variable replaces two or more dependent variables, it is evident that estimates of statistical power can proceed in a manner analogous to that used in the univariate approach.

For the two-treatment-group case, Hotelling (1931) derived the statistic T^2, which is a generalization of Student's t. T^2 relates exactly to the F distribution as follows.

1. Matched samples or subject as own control:

$$\frac{N - P}{(N - 1)P} T^2 = F \qquad df \langle \begin{matrix} P \\ N - P \end{matrix}$$

2. Two independent groups:

$$\frac{n_1 + n_2 - P - 1}{(n_1 + n_2 - 2)\, P}\, T^2 = F \qquad df \left\langle \begin{matrix} P \\ n_1 + n_2 - P - 1 \end{matrix} \right.$$

Where N = number of matched pairs, P = number of dependent variables, df = degrees of freedom, and n_i = number of subjects in one treatment group.

Due to the mathematical relationship between F and Hotelling's T^2, the procedure outlined for univariate power analysis is applicable, with slight changes, to MANOVA power estimation.

Just as for ANOVA, power estimation in MANOVA requires the calculation of phi.

$\phi = [\delta^2/(P + 1)]^{\frac{1}{2}}$

P = number of dependent variables

δ^2 = Hotelling's T^2

Computation of Hotelling's T^2 will directly yield the value of delta-square. Otherwise, delta-square can be easily determined from computer output of a MANOVA program. The eigenvalue resulting from factoring the product matrix ($SSCP_{error}^{-1} \cdot SSCP_{treatment}$) is simply multiplied by the error degrees of freedom (for a single univariate variable) to yield Hotelling's T^2 or delta-square. The calculated phi value is located in the appropriate table or chart, which is determined by the degrees of freedom for the particular F ratio associated with Hotelling's T^2 for either independent groups or matched pairs (or subject as his or her own control). Power of the statistical test is read from the table or chart in precisely the same manner as for univariate power estimation.

The procedure just outlined for two-group power estimation in MANOVA may be utilized to estimate size of sample required in the same experiment or in a subsequent one in order to attain the desired statistical power. Both the eigenvalue and error degrees of freedom are used for this purpose. Essentially, one inserts increasingly larger error degrees of freedom into the equation ($\delta^2 = df_e\, \lambda$) until the computed phi value (associated with increased error degrees of freedom) attains the desired level of statistical power. Then the error degrees of freedom are converted to sample size. The eigenvalue in the equation may be obtained from the study one made that resulted in retention of the null hypothesis, from a pilot study, or from some average or typical value reflecting prior published research. In chapter 5, an example is given of sample size estimation to attain a stated power level for a two-treatment-group matched pairs design.

If more than two treatments are used (and more than two dependent variables) in a MANOVA design, estimation of statistical power is far more difficult than for the two-group case. Two or more eigenvalues result from the MANOVA analysis, and each eigenvalue is associated with an independent synthetic variable. Stevens (1980) presents a lucid treatment of the problem together with helpful power tables and worked examples for estimating power in more complex (but independent) MANOVA designs. Stevens's tables are entered via a simple sum of the eigenvalues. Statistical power estimates for each of the four MANOVA criteria (Wilks's lambda, Roy's GCR, Lawley–Hotelling trace, and Pillai's trace) are available in the tables. The tables provide for number of dependent variables, sample sizes, and number of treatment groups. The tables allow the investigator to estimate (roughly) the statistical power for research just completed or, if pilot data or prior reported data are used, to determine the size of the sample required to attain the desired statistical power.

Bonferroni *t*

The following procedure enables an investigator to maintain a desired alpha level for a set of comparisons while testing significance between pairs of treatment means on one dependent variable at a time. The Bonferroni *t* is perhaps the simplest statistical procedure that may be appropriately applied to the multiple comparisons possible in MANOVA data. As we indicated earlier, the procedure involves dividing up the alpha value according to the (set) number of comparisons to be made; the resulting quotient is used to indicate the alpha level required before any single multiple-comparison *t* test in the set can be significant. The consequence is that the alpha level is appropriately controlled for the set, or family, of comparisons. The family, or set, of comparisons must be formulated prior to examining the data. If the number of multiple comparisons to be made is quite small, the statistical power of the Bonferroni method exceeds that of other alternative methods. But aside from its simplicity of use and superior power when few comparisons are made, there seems to be little to recommend it in a MANOVA situation.

A number of serious disadvantages attend the use of the Bonferroni *t* in a MANOVA situation. First, assuming the usual ANOVA design with a rather large number of potential comparisons, statistical power is small. Furthermore, if the investigator selects a small set of comparisons in order to gain power, he or she obviously discards a large portion of the research design, thereby calling into question the justification for employing the larger design in the first place. However, before we deliver a final statement on this point, we should note that this weakness of the Bonferroni approach may be alleviated by a redefinition of the set, or family, of comparisons. For example, Tukey (1953) and Myers (1979) conceive of a set of comparisons as corresponding to a main, or treatment, effect in ANOVA. Following this logic, several independent sets are formed in a complex ANOVA design. This procedure results in an increase in the power of the test in a typical ANOVA or MANOVA design.

The second and perhaps most important drawback to the use of the Bonferroni t in a MANOVA situation is that no check is made on the dimensionality of the dependent variables, and consequently the nature of the synthetic variable that is developed in MANOVA is not available for examination. As the reader will note from chapter 5, use of variables to measure different dimensions, based on the assumption that "different" labels imply different dimensions, is a very risky gamble. Obviously a variable that plays an important suppressor role in forming a synthetic variable will appear to be of no importance in the Bonferroni approach, and its contribution will be overlooked completely.

Classic MANOVA Procedure

Two very different methods have been associated with classic MANOVA procedures. The first method focuses attention on synthetic variables, whereas the second method deals exclusively with the original dependent variables.

In the first procedure, the dimensionality of the discriminatory matrix $(SSCP_{error}^{-1} \cdot SSCP_{treatment})$ is determined. Then synthetic variables are derived, along which the treatment groups are maximally separated with minimal error variance. A MANOVA test follows and appropriately takes into account the statistical and substantive significance of separation of the treatment groups along the synthetic dimension. Roy's GCR test takes into account only the first dimension, whereas the other MANOVA tests (Lawley–Hotelling trace, Wilks's lambda, and Pillai's trace) take into account all dimensions simultaneously. If the chosen MANOVA test is significant, the separation of treatment-group centroids along each synthetic variable is tested for statistical significance.

The second procedure tests the statistical significance of difference between means of treatment groups on one dependent variable at a time, although it is possible to compare one treatment group on more than one dependent variable. The procedure holds Type I error at or below the chosen level and is regarded as a very conservative test. These comparisons are termed simultaneous linear contrasts (Berger, 1978). One major advantage of the method is that differences between treatment group means required for significance (on a single dependent variable) may be readily calculated with only the need for extant multivariate tables (see Timm, 1975). A major disadvantage is that synthetic variables are ignored. The method will now be described in some detail.

Recall that Student's t test can be reexpressed so as to determine the size of the difference between means required for significance at a desired alpha level.

$$\text{Critical difference } (CD) = (M_1 - M_2) = t_\alpha \sqrt{ms_e \left(1/n_1 + 1/n_2\right)}$$

t_α = Student's t ratio required for significance at the selected alpha level

n_i = number of subjects in treatment group

ms_e = error term in denominator of F ratio

For simultaneous linear contrasts, a value K is computed from the MANOVA test being used and K is substituted in place of the value t_α in the critical difference formula. A MANOVA table is used to determine the magnitude of the test statistic (Wilks's lambda, Roy's GCR, Lawley–Hotelling trace, or Pillai's trace) required for significance at the chosen alpha level. The tabled value is substituted into the appropriate formula to determine K.

Wilks's Λ

$$K = \left[df_e \left(\frac{1 - \Lambda}{\Lambda} \right) \right]^{1/2}$$

Pillai's V

$$K = \left[df_e \left(\frac{V}{1 - V} \right) \right]^{1/2}$$

Roy's θ

$$K = \left[df_e \left(\frac{\theta}{1 - \theta} \right) \right]^{1/2}$$

Lawley–Hotelling τ

$$K = (df_e \ \tau)^{1/2}$$

In each of the preceding equations for determining K, df_e refers to the degrees of freedom associated with the error term for the particular dependent variable being examined. The calculated value of K is inserted into the formula for the critical difference, and the solution of the equation yields the size of the difference between treatment means required for significance on that one dependent variable. Determination of the critical difference for a second dependent variable requires only the insertion of the mean square for error (ms_e) of the second dependent variable into the same equation.

The parameters for entering tables of Roy's GCR, Lawley–Hotelling trace, and Pillai's trace are S, M, and N, which are defined in chapter 2. The table for Wilks's lambda is entered with the following parameters: p = number of dependent variables, $q = df_{treatment}$, and $n = df_{error}$.

Hummel–Sligo Procedure

Hummel and Sligo (1971) studied a strategy for analyzing MANOVA data (Cramer & Bock, 1966) that appears to be superior to other classic procedures. Using computer simulations, the investigators confirmed the highly conservative nature of the classic MANOVA procedure. However, they demonstrated that routine use of ANOVA for each dependent variable, following a significant overall MANOVA, resulted in probability values that were close to the desired alpha levels.

This finding is good news to the researcher whose research training has been confined to ANOVA. It provides a rationale whereby the investigator may legitimately analyze each of the separate dependent variables that are measured in the same ANOVA design. It is necessary that the investigator perform preliminary MANOVA tests. Assuming a significant MANOVA test, the investigator is free to apply ANOVA to the separate dependent variables with the assurance that the comparisons are controlled approximately at the chosen alpha level. MANOVA then provides a preliminary test to ANOVA much as the F ratio provides a preliminary test to t.

It should be noted that overall MANOVA tests are performed for each of the interaction and main effects inherent in the ANOVA design. The same strictures that apply to interpretation of ANOVA main and interaction effects apply to the MANOVA procedure. For example, if the MANOVA test for interaction is significant, then one may examine the appropriate cells (simple effects) in ANOVA fashion for each dependent variable. However, main effects included in the interaction are no longer of interest.

The Hummel–Sligo procedure, like the Bonferroni t, ignores the value inherent in examining the synthetic variables. It is, though, simpler to use and more powerful than simultaneous linear contrasts.

Mixed Strategy

In addition to the advantage of permitting statistical significance tests, synthetic variables are potentially important in summarizing and interpreting multivariate research data. For this reason the mixed strategy accords equal interest to both the synthetic variable(s) created by MANOVA and the original dependent variables. The approach is essentially the same as the Hummel–Sligo method except that the synthetic variable(s) are not slighted. After a preliminary MANOVA test has been performed, the synthetic variables(s), on which the treatment groups are significantly separated, are examined in order to test hypotheses regarding the rank and nature of the treatment effects on the dependent variables. Then a serious attempt is made to interpret the synthetic variables in terms of the pattern of loadings of the variables on the factors. The contribution of individual variables to the synthetic variables is highlighted.

Attention then shifts to ANOVA results for the separate dependent variables. Findings at this level may be considered jointly with those at the synthetic variable level to provide a highly informative indication of the outcome of the research.

Attempts to interpret the synthetic variable(s) should take into account the fact that the synthetic scores derive from a principal-components factor analysis. There are a number of distinct advantages in the principal-components solution. First, each factor is extracted in a sequence such that a maximum of variance is accounted for. Second, the nature of the extraction process permits a sophisticated test of statistical significance of each factor (root). In addition to these desirable properties of the principal-components solution, the synthetic scores may frequently be interpreted easily and meaningfully in terms of the original variables.

If the synthetic variables are meaningful as well as statistically significant, the investigator will likely wish to compare the treatment groups on the synthetic variable(s). In order to determine whether separations between the treatment groups on the synthetic variable are statistically significant, the researcher must consider two major issues. The first concerns the nature of the comparison to be made, whether the comparison represents an a priori, orthogonal contrast or an unplanned (data-snooping) comparison. The very same issue arises in univariate ANOVA comparisons.

The second issue concerns the weights that are used to form the synthetic variable, whether they are suggested by prior research and theory (e.g., use of factor scores) or are derived (post hoc) from the data that is being analyzed. Figure 3–1 clarifies the four different conditions that result from joint consideration of the nature of the contrast to be made and the method of deriving the discriminant weights.

Origin of Weights
for
Determining Synthetic Score

	A Priori	Post Hoc
A Priori	t (1)	Hotelling T^2 (2)
Post Hoc	Scheffé t (3)	Multivariate Extension of Scheffé t (4)

Nature of Contrast (row label)

Figure 3–1. Decision Chart for Selection of Appropriate Statistical Test

Cell 1 involves planned a priori comparisons and a priori weights for forming the synthetic variable. Under these circumstances the conventional t test used in ANOVA may be applied to the synthetic scores. In cell 3 contrasts are not planned in advance, but the weights for forming the synthetic variable are selected a priori. The Scheffé t is appropriate for this situation. In cell 2 the contrasts are selected a priori, but the weights for determining the synthetic score are derived from the data being analyzed. Hotelling's generalization of the univariate t (T^2) is appropriate under these circumstances. Cell 4 involves post hoc contrasts and post hoc weights for the synthetic variable. A generalization of the Scheffé t test to a multivariate equivalent is available. The data analyst has several options in deriving the generalized statistic. These options arise because of the existence of four classic multivariate test criteria and the frequent extraction of more than one eigenvalue. Harris (1975) favors the use of Roy's greatest characteristic root, which uses only the first eigenvalue in deriving the test. We favor the use of the Lawley–Hotelling trace because of its simplicity, because it uses all of the extracted eigenvalues, and because its sampling distribution is unaffected by the form of distribution of the trace in the eigenvalues.

The statistical tests under each of the four conditions (Figure 3–1) can be expedited by determining the critical difference between group means on a synthetic variable (centroids) required for significance at the chosen alpha level. The critical difference required for significance for cells 1 and 3 can be determined by rearranging the formula for t.

$$t = (M_1 - M_2)/S_{\text{diff}_m}$$
$$(M_1 - M_2) = t_\alpha \cdot S_{\text{diff}_m}$$
$$CD = t_\alpha \cdot S_{\text{diff}_m}$$

For statistical contrasts represented by cells 1 and 3 (Figure 3–1), CD between centroids requires the tabled values of Student's t (or its change to a Scheffé equivalent) to be multiplied by the standard error of the difference between means. The MANOVA computer program by Barker and Barker (1977) scales each synthetic score such that the error source variance is 1.00, thereby simplifying the calculations.

For cells 2 and 4 in Figure 3–1, the component $(df_e \tau_\alpha)^{1/2}$ is substituted in place of t_α. The Lawley–Hotelling trace (τ) required for significance at the chosen alpha level is multiplied by the error degrees of freedom for a single univariate dependent variable. Then the root of the product is obtained and is entered into the equation for the critical difference. If treatment groups contain an equal number of observations, critical differences (at the .05 and .01 alpha levels) for each of the four cells are computed and are printed routinely by the Barker and Barker (1977) MANOVA program.

4

Classic Research Designs

Although there are numerous good texts on experimental design within the context of analysis of variance, few emphasize overall organizational principles. This latter concern seems especially important in a text that stresses statistics as a tool for decision making.

The reader, whether a novice or sophisticate in experimental design, should find this chapter beneficial. The sophisticate will profit from a brief but unusual and extensive review of experimental design principles; both sophisticate and novice will be provided with a standard framework and nomenclature for viewing all research designs.

Certain basic issues are common to all research designs. These include the relationship between ANOVA and MANOVA and the distinction between a study and an experiment, matters that will be discussed at some length. In conclusion we shall show how all classic research designs evolve from the two classic t tests.

Two Preliminary Issues

Basic Distinction between ANOVA and MANOVA

The primary and perhaps the sole design distinction between ANOVA and MANOVA is that ANOVA is restricted to a single dependent variable, whereas MANOVA employs simultaneously two or more dependent variables. Either ANOVA or MANOVA may employ any number of independent variables. In view of the structural similarity between ANOVA and MANOVA, classic research designs are applicable to both and are therefore reviewed together in this chapter. Apart from the design features, both ANOVA and MANOVA similarly employ statistical tests of significance and substantiveness. However, in the case of MANOVA, these tests are considerably more complex and usually require a computer for solution.

Study versus Experiment

A basic feature of all ANOVA and MANOVA designs is that they are neutral with respect to the origin of the data. This very basic issue deserves further discussion. The ANOVA or MANOVA structure "works" (at the mathematical level) appropriately, whether it is applied to random numbers, social security numbers, or data resulting from a study or experiment (Lord, 1953). Although the latter point seems obvious, the use of ANOVA (or MANOVA) for observational data is sometimes inappropriately questioned.

Granted that the ANOVA (or MANOVA) model may be applied to either experimental data or study data, the interpretation of the statistical results normally requires greater care in the case of a study than in the case of an experiment. As we noted in chapter 1, the primary distinction between a study and an experiment hinges on control of variables. Whereas in an experiment results may be readily attributed to variables manipulated by the experimenter (other variables being under control), in a study results may be due to pseudoindependent variables or other variables not controlled and generally not known.

Control Checklist

To bring variables under control in a research investigation, it is helpful to have a standard checklist of the kinds that require control. A checklist by Lindquist (1953) is simple and inclusive and is valuable in the preliminary research design stage or in the postdesign stage when one is considering variables that may have influenced the results but were not expected to do so. Variables or errors are viewed as falling into one of three types: S, G, and R. Type S refers to "subject" variables including such differences between people as age, height, IQ, attitude, and so forth. Type G variables influence the members of one or more treatment groups but are not available (or are not available to the same degree) to influence all treatment groups equally. For example, research involving members of one experimental group who are tested at night and members of other experimental groups who are tested during the day may be influenced by the time when testing occurred. As another example, assume that a member of one treatment group undergoes a grand mal epileptic seizure during testing of the group. The fright that the incident produces in other members of the group may affect the results for this treatment group, whereas the other treatment groups, not having been exposed to the seizure, are not affected. An actual instance of a Type G error that went unnoticed in behavioral research for many years involved the association of male and female experimenters with different treatment groups. Due to the considerable influence of this variable found in several studies, it is now frequently used as an independent variable in its own right.

Type R error as used by Lindquist refers to "regional" variables that produce real but unequal impact on an experiment. For example, an experiment comparing a conventional and "new" method of teaching mathematics may produce in a university laboratory school results very different from those obtained in a school in a rural, isolated, impoverished setting. The effect of the teaching methods may be "correct" for each setting, but the effects are different for the two settings. For research purposes the term *regional* may easily be generalized to (genetically different) litters of animals, culturally different groups (although geographically located in proximity), and so forth.

It is, of course, difficult to classify with certainty some variables in a given experiment. Fortunately it is not necessary to do so in order to profit from use of the error checklist. It is sufficient to recognize and then to be able to take into account (or to control) the variables identified by the scheme.

Occasionally, variables may be brought under direct control; for example, room temperature and humidity, time of day for testing, and so on. When it is not possible to control variables directly, control may be exercised by randomizing. Random assignment of subjects to treatments is now a standard technique for controlling Type S variables. An unexpected benefit accrues from the randomization procedure. Treatment groups are found to be roughly comparable on variables recognized as being important to the investigation and also on variables about whose importance the investigator is uncertain. Randomization may be employed also to control for Types G and R errors, but they must first be identified. Sometimes it is possible to control numerous Type G influences by testing all subjects in the same setting.

Perhaps the more common usage of randomization should be mentioned. By randomly selecting a pool of subjects from a defined population, results of the research can be appropriately generalized to the population.

Origin of All Classic ANOVA Designs

All classic ANOVA designs may be viewed as originating from two basic types of t tests: the t test for independent groups and the t test for matched pairs (or for the subject as his or her own control). Consider the computational formulas for the two techniques:

1. t test for independent groups

$$t = (M_1 - M_2) / \sqrt{S_{M_1}^2 + S_{M_2}^2}$$

2. t test for matched pairs (subject as own control)

$$t = (M_1 - M_2) / \sqrt{S_{M_1}^2 + S_{M_2}^2 - 2r_{12} S_{M_1} S_{M_2}}$$

The two formulas are identical except for a subtraction component in the divisor of the t test for matched pairs. The subtraction component results from pairing of subjects such that correlation may be computed. Since the investigator produces the correlation deliberately, as a means of gaining control, the term may be deducted from the remaining error components. Note that the larger the correlation, the greater the subtraction term and the smaller the error-term divisor. This feature has the effect of enlarging the t ratio, thereby increasing the statistical power of the test; that is, the probability is increased of rejecting the null hypothesis when it is false.

The correlation term does not appear in the formula for the t test for independent groups because there is no rational basis for pairing the different subjects in different groups in order to compute a correlation. Another way of viewing the absence of the correlation term involves noting that the correlation between random pairs would average out to zero, thus canceling the subtraction term. Figure 4–1a and 4–1d depict the similar design features of the two t tests: Each involves one independent variable (two levels) and one dependent variable.

One class of ANOVA designs results from the t test for independent groups; a second class of designs emanates from the t tests for matched pairs. A third class of designs is the result of mixing designs from the first two classes. Examples of classic designs from each of the three classes will be given. The manner in which the illustrated designs control (or fail to control) for the three types of error (S, G, and R) will be highlighted.

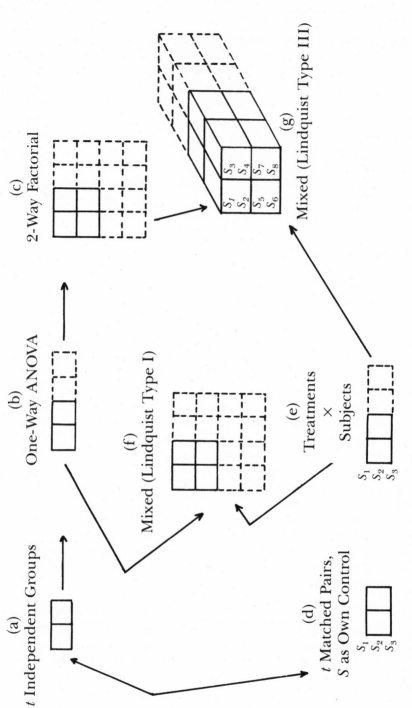

Figure 4–1. The Origin of All Research Designs in the Two Classic t Tests

Extension of *t* Test for Independent Groups

Simple "Randomized" Design (One-Way ANOVA)

Figure 4–1b depicts the simple randomized (one-way ANOVA) design. It simply employs additional treatment groups (levels) beyond the two that compose the *t* test for independent groups. The designation *randomized* in the title highlights the attempt at control of variables characteristic of an experiment. The design provides for control of S variables by randomizing subjects to treatment groups; however, no provision is made for errors of Types G and R, and results must be carefully qualified with that in mind. Errors of Types G and R, if present, will be confounded with treatment effects and cannot be extricated.

If subjects are not randomly assigned to treatment groups, the design is more accurately termed a one-way ANOVA (study). The investigator must be prepared to qualify results as possibly being due to the treatments or one or more of the three types of error.

Two-Way Factorial

In addition to having two or more levels of one independent variable, the two-way factorial design makes provision for two or more levels of a second independent variable. Figure 4–1c depicts the design. The design provides for control of Type S error by randomizing subjects to all treatment cells (with the restriction that all cells contain an equal number of subjects). Known Type G error sources can be taken into account (controlled) for main effects of row and column treatments by randomizing all known sources of Type G influence across treatment cells. For example, time of day of testing might be randomized across treatment cells such that this Type G influence would exert no systematic effect on any main row or column treatment effect.

Although Type G influence can be controlled for main effects in the factorial design, such is not the case for interaction effects. It is helpful to denote a true treatment interaction as *intrinsic* and an interaction resulting from an unknown Type G influence as *extrinsic* (Lindquist, 1953). The decision as to which of these two types of interaction applies in a given instance must remain subjective on the part of the investigator. In a study where no control of variables is attempted, all effects—main and interaction—may result from Types S, G, or R error.

N-*Way Factorial*

Theoretically no restriction is placed on either the number of independent variables for factorial designs or the number of levels for a particular independent variable. The same considerations regarding Types S, G, and R errors that were discussed for two-way factorial apply for *N*-way factorial.

Extension of the *t* Test for Matched Pairs (Subject as His or Her Own Control)

Before presenting an ANOVA design that results from an extension of the *t* test for matched pairs, it is useful to consider three methods whereby subjects may be matched.

Exact Method

For a set of *k* treatment groups, a set of *k* subjects who are matched precisely on relevant variables is selected. Members of the matched set are randomly assigned, one to each treatment group. The procedure is repeated until the subject pool is used up. The more exact the matching criteria employed and the greater the number of matching variables, the more difficult it is to achieve matched groups. Severe problems may then attend attempts at generalizing research results from such a highly select sample.

Ordering Method

The subjects are rank ordered in terms of their measurement on some appropriate control variable(s). Then the first *k* rank-ordered individuals are each randomly assigned to the *k* treatments. The next *k* rank-ordered subjects are randomly assigned, and so forth. By using all subjects in the sample, this method avoids the problems of generalizing results that plague the so-called exact method.

Subject Used as His or Her Own Control

The rationale underlying this method is that a subject is a more perfect match for himself (or herself) than any other individual. However, the method's simplicity must be weighed against some important problems that arise from its use. These problems principally result from the influence of order and sequence on the treatment effects.

Order and sequence may be controlled by randomization and may thus confound their effects with treatments in an unbiased manner. On the other hand, order and sequence may be systematically manipulated by the investigator. They then attain the status of independent variables, and provision must be made for their evaluation in a more complex ANOVA design.

Treatments × Subjects Design (Randomized Blocks)

Figure 4–1e depicts this type of design. It involves simply increasing the number of treatment conditions beyond the two of the *t* test. As indicated earlier, either subjects are matched, or each subject is used as his or her own control. The design consists of two or more treatment levels of one independent variable and one dependent variable. The subject pool is randomly selected from a defined population. In the event of the use of matched subjects, each set of *k* subjects is randomly assigned to the treatments. As the alternate label (*randomized blocks*) implies, the unit may consist of different litters of animals, in which case random assignment of individuals to treatments from litters is accomplished. The error term for this type of design consists of the interaction between subjects and treatments.

Mixed Designs

The term *mixed* denotes an ANOVA design that results from combining variations of the *t* test for independent groups with variations of the *t* test for matched pairs. The resulting ANOVA design contains both between- and within-subjects (or matched pairs) effects. These effects are analyzed separately. The analysis essentially consists of two steps. First, a total score is obtained for each subject (or *k* matched subjects) by summing across the repeated measures. The design is thereby simplified, enabling the totals to be analyzed in typical *between-subjects* fashion. Second, the repeated measures (or matched measures) are directly analyzed in *within-subjects* fashion. The two analyses are reported and considered separately.

The two stages of the mixed design require two different error terms to test treatment effects. Each of the error terms represents an average of variability contained in particular units. The units in the between-subjects designs are normally located in the individual treatment cells,

whereas in the within-subjects designs they normally comprise the Treatments × Subjects interaction pooled across separate Treatments × Subjects designs. Tests of homogeneity of variance of the error sources are
applied to the appropriate units. Other complications of within-subjects
designs are discussed in chapter 6.

Although a wide variety of mixed designs can be constructed, only two
classic types will be described. These two examples provide the bulk of
research usage; those interested in other varieties should consult Winer
(1962), Kirk (1968), Myers (1966), and Lindquist (1953).

Lindquist Type I (One Between- and One Within-Subjects)

Recall that the first step in analyzing a repeated-measures design is to
sum the repeated measures. When this has been done, the design reduces
to a simple randomized design and is so analyzed (See Figure 4–1f). The
next step involves viewing the design in terms of its repeated measures. It
is clear, then, that the design consists of replications of the Treatments ×
Subjects design—one replication for each level of one of the independent
variables.

The simple randomized aspect of the design provides for control of
Type S variables. However, Types G and R variables are not controlled and
must be evaluated subjectively by the investigator. The Treatments ×
Subjects aspect of the design provides for further Type S control and for
greater statistical power in that individual differences between subjects are
isolated and do not appear in the within-subjects error term.

This design is useful for a situation in which a comparison between a
pretest and a posttest is needed for a control and for one or more
treatment groups. In the event of just two treatment groups and two
repeated measures, the design incorporates as its simplest case an older
classic procedure (Walker & Lev, 1953). The difference between pretest
and posttest is determined for each individual, so that the design reduces
to two sets of independent difference (or change) scores. A t test is then
computed on the difference between mean difference or change scores. If
the same data are analyzed by the Lindquist Type I design, the identical
test is found in the within-subjects test of interaction between pretests and
posttests and the treatment groups.

Lindquist Type III (Two Between- and One Within-Subjects)

From Figure 4–1g note that the design is constituted from the two-way
factorial design and Treatments × Subjects design. After the preliminary

step of summing the repeated measures for each individual, the design is reduced to a two-way factorial design consisting of subject totals. Then, on considering the repeated measures, the design replicates the Treatments × Subjects design—one replication for each treatment cell. Type S error is controlled by random assignment of subjects to the treatment cells, and greater statistical power is afforded by the within-subjects portion of the design. By randomization, Type G errors can be controlled for main effects but not for interactions. The design makes no provision for Type R errors.

5

Applications of MANOVA to Classic Research Designs

Preliminary Considerations

It seems appropriate at this point to introduce the reader to several practical matters not yet discussed and to recall several topics previously developed.

MANOVA Computer Programs

Several MANOVA programs are currently available: SAS (Barr et al., 1979), BIOMED (Dixon, 1979), Multivariance (Finn, 1976), SPSS (Nie et al., 1975). They differ in many respects including the mathematical approach to solution, ease of use, and practical decision value of output.

With the recent updated versions of MANOVA (SPSS) and BIOMED (BMDP P4V), all of the current statistical packages provide in their printout the classic MANOVA tests (BMDP P4V omits Pillai's trace), the univariate F tests on the dependent variables, discriminant weights, and correlations between dependent variables and each synthetic variable (SPSS does not report whether the error source or total variation is used in the calculation). See chapter 2 for a discussion of why the error source should be used in computing correlation between dependent variables and synthetic scores.

Users of the updated SPSS MANOVA program may request a factor analysis of the intercorrelation matrix (based on pooled error sources) and may select one of three methods of rotation. This option is valuable for examining the dimensionality (factor structure) of the dependent variables (unaffected by the treatments). It is assumed that the error source matrices are homogeneous, and this assumption is tested by the Box test.

Both updated versions of SPSS MANOVA and BIOMED (BMDP P4V) provide for direct MANOVA and ANOVA treatment of repeated measures. None of the statistical packages (except Barker & Barker, mentioned later) provide statistical significance tests for difference between pairs of centroids on a synthetic variable.

Users of the updated versions of SPSS MANOVA and BIOMED (BMDP P4V) may experience considerable difficulty in learning to employ them properly because they are complex, general-purpose programs. BIOMED update cautions that full usage of the MANOVA program may require a considerable background in ANOVA and that other related technical publications should be consulted. The reader is reminded that this text uses several sample problems from Timm (1975). The raw data are given, and the reader is urged at a minimum to analyze these data using an available computer program and to check the accuracy and labeling of the results carefully.

The computer program used in solving the examples given later in this chapter is identified as SPEC78 (Barker & Barker, 1977), one of numerous statistical programs composing the Behavioral Sciences Statistics Program Library prepared by the authors. All computer programs in the library (including SPEC78) incorporate principles of human factors engineering to reduce usage errors; the programs feature as output whatever is needed for practical, scientific decision making. The SPEC78 program is completely applicable to both ANOVA and MANOVA data and thus makes available in one program all of the classic designs considered in chapter 4. For unusual types of designs, or designs involving four or more dimensions, the user must furnish sums-of-squares-and-cross-products matrices. SPEC78 is available on request from the Seebeck Computer Center, The University of Alabama.

Some Principles of MANOVA

One principle of MANOVA is so basic that it should be carefully considered in the design stage of any MANOVA research. This principle applies to the number of factors (dimensions) that can separate the group centroids (means in the factor space): The number of factors that can be extracted is limited to the number of dependent variables or the number of degrees of freedom for the treatment under consideration, whichever is smaller.

An example of the application of the principle follows. Suppose an investigator takes as dependent variables two different measures of each of three supposedly different dimensions that are expected to separate the treatment groups. A simple randomized design is used involving three treatment groups. Thus the maximum number of factors that can be extracted is two (three treatment groups minus 1.0). The investigator is therefore unable to test separation of treatment groups along the hypothesized three dimensions. In order to confirm the three-dimensional nature of the group separations, the investigator should have employed a minimum of four treatment groups.

A second principle of MANOVA relates to an earlier discussion regarding the differential sensitivity of the various MANOVA test criteria to the distribution of trace (Barker & Barker, 1979). The investigator is urged to consider the following points in the design stage of MANOVA research.

1. If multiple dependent variables are used in an effort to gain a more reliable measure of a single dimension, it is anticipated that most of the trace will load on the first root, and Roy's greatest characteristic root criterion should be employed as the test statistic of choice.

2. If a priori factor pure tests or uncorrelated factor scores derived for the occasion are used as dependent variables, the trace is expected to distribute relatively across the roots, and Wilks's lambda statistic is expected to be most sensitive.

3. Aside from the two just mentioned extreme distributions of trace, the four MANOVA test criteria (including Lawley–Hotelling trace and Pillai's trace) manifest little in the way of differential sensitivity. However, Olson (1976) argues for general superiority of Pillai's trace criterion, especially where heterogeneity of error dispersion matrices is present. Stevens (1979) challenged the general superiority of Pillai's trace but acknowledged its superiority under conditions of extreme heterogeneity of dispersion matrices.

4. Ideally, an investigator should select and interpret a test statistic in MANOVA on an a priori basis. However, due to the varying sensitivity of the test criteria to the manner of distribution of trace, it seems desirable to report the results of all four.

The third principle applies to the strategy governing the interpretation of interaction and main effects. Decision strategy in ANOVA permits the examination of main effects of treatments only if interactions involving the treatments are nonsignificant. If interaction is significant, then the appropriate treatment cells (simple effects) require analysis. This same decision strategy applies to MANOVA (Morrison, 1967).

Origin of Data in Examples

The data used to illustrate the classic MANOVA analyses were carefully selected from Timm (1975), available master's theses and doctoral dissertations, and other actual research studies. For this reason, it is possible to illustrate both desirable and undesirable practices in the design and analysis of MANOVA research. Furthermore, although some of the reports qualify as experiments, others have the status of studies, thereby permitting a demonstration of the greater need for qualification of findings in studies. The examples from Timm (1975), which feature matrix algebra solutions, allow the reader to acquire greater mathematical sophistication by coordinating the current text with Timm's mathematical presentation.

Procedure to Be Followed for Each Classic Design

For each classic design, we provide a research report similar to a journal presentation. The report includes a statement of the problem, a succinct report on method and procedure, a summary of computer data analysis, and alternate interpretations of the results. We hope that readers will feel that they are "looking over our shoulders" as we view the research problem and analyze it. We mean to stress desirable and undesirable practices and their consequences.

Classic Designs

Simple Randomized Design (One-Way MANOVA)

Problem An exploratory study was conducted to determine if male university students differ in attitude ratings on 10 miscellaneous topics according to major field of study.

Method Data were accumulated from students in elementary statistics courses over several years. Subjects selected for inclusion in the study were 112 male students. They represented four major fields of study as follows:

1. educational psychology ($n = 19$)
2. speech ($n = 16$)
3. criminal justice ($n = 60$)
4. commerce and business administration ($n = 17$)

Students expressed their attitudes by rating each of 10 topics on a 5-point scale:

1. extremely dislike
2. dislike
3. neutral
4. like
5. extremely like

The miscellaneous topics were:

1. mathematics
2. statistics
3. English literature
4. football
5. basketball
6. science
7. rock music
8. classic music
9. country music
10. marijuana

Results Although three tests of homogeneity of dispersion matrices are available in the computer printout, only the most versatile (Box's F test) is reported here. Box's F test yielded an F value of 1.39, which with 165 and 7553 degrees of freedom is significant at less than the .01 level. This finding indicates lack of homogeneity of variance–covariance matrices, which are pooled to form the error term used in MANOVA. An examination of the determinants of the indicated error sources (see Table 5–1) suggests Group 3 (criminal justice) as an outlier. As in ANOVA, an outlier error source serves to bias the error term in the direction of the outlier group. In this instance, a conservative MANOVA test results.

A one-way multivariate analysis of variance produced four MANOVA test criteria, which are displayed in Table 5–2. Wilks's lambda (.565) is typically converted to an F ratio in order to test for statistical significance. The resulting F is 4.19, which with 30 and 585 degrees of freedom is significant at less than the .001 level. The other three test criteria, Roy's

Table 5–1. Determinants from Within-Groups Error Sources

Group	Determinant
Educational psychology	.155
Speech	.001
Criminal justice	.338
Commerce	.010

Table 5–2. Four Test Criteria Resulting from MANOVA

		Table entry point		
	Statistic	S	M	N
Wilks's lambda	.565***	—	—	—
Lawley–Hotelling trace	.685**	3	3	98.5
Roy's greatest characteristic root	.350**	3	3	98.5
Pillai's trace	.484**	3	3	98.5

*$p < .05$. ** $p < .01$. *** $p < .001$.

GCR, Pillai's trace, and the Lawley–Hotelling trace, are typically referred to special multivariate tables to test for significance (Timm, 1975).

In each instance, the tables are entered according to three parameters: S, M, and N. These values are given in Table 5–2. A check on these three test criteria indicates that each is significant beyond the .01 level. Thus all four multivariate test criteria agree, indicating that the four groups of university students differ significantly in terms of attitude ratings on miscellaneous topics.

Three roots or dimensions were extracted by the MANOVA program. A chi-square test (illustrated in the appendix) performed on each of the three roots resulted in statistical significance for the first two, as shown in Table 5–3. Table 5–4 presents intercorrelations between the synthetic (discriminant) and dependent variables. These correlations are similar to factor loads (in factor analysis) and permit one to infer the nature or meaning of the dimensions.

Following convention, a variable loading (correlation) of $\pm.3$ or greater on a factor is required in order to consider the variable as "defining" the factor. Variables mathematics, statistics, football (negative loading), and science define the first synthetic variable. This might be considered a *quantitative* dimension. The second synthetic variable correlates with En-

Table 5–3. Statistical Tests of Three Extracted Roots

Root	% Trace	χ^2	df
1	78.47	87.74***	12
2	16.44	21.76*	10
3	5.10	7.00	8

*$p < .05$. **$p < .01$. ***$p < .001$.

Table 5–4. Intercorrelations between Synthetic (Discriminant) and Dependent Variables

	Synthetic variable		
Variable	I	II	III
Mathematics	.49	.17	−.24
Statistics	.71	.10	.00
English literature	.06	.79	.28
Football	−.31	.05	.02
Basketball	−.20	−.18	.17
Science	.53	.05	.07
Rock music	.02	.07	.36
Classic music	.15	.35	−.18
Country music	−.15	−.26	.34
Marijuana	.24	.11	.73

glish literature and classic music and might be termed a *humanistic* factor. The third synthetic variable correlates positively with rock music, country music, and marijuana and might be termed a *protest–unconventional* dimension. Of course, this third dimension does not significantly separate the four groups and ought therefore to be interpreted tentatively.

Table 5–5 displays mean scores (centroids) on the synthetic variables for each of the four groups. The centroids are expressed in standard units so that differences between group centroids can be evaluated as standard deviates. We shall rank order the means for the four groups on the first synthetic dimension (which we viewed as quantitative). Educational psychology has the highest mean (2.92), followed by commerce (2.16), then speech (1.53), and criminal justice (1.33).

On the second dimension, interpreted as humanistic, rank ordering gives speech majors the highest mean (3.67), followed by commerce (2.57),

Table 5–5. Group Centroids on the Three Synthetic Variables

	Synthetic variable		
Group	I	II	III
Educational psychology	2.92	2.55	2.81
Speech	1.54	3.67	2.82
Criminal justice	1.33	2.32	2.83
Commerce	2.16	2.57	2.14

educational psychology (2.55), and criminal justice (2.32). On the third dimension, group means are very close, and in view of the lack of significant separation, no interpretation is made.

To this point, we have viewed the synthetic variables that result from the application of MANOVA to the data. An alternative approach is to employ MANOVA as an omnibus preliminary test to determine whether it is plausible to examine the separate dependent variables in univariate fashion (Hummel & Sligo, 1971). In view of the significant MANOVA tests, the Hummel–Sligo strategy permits univariate F tests on the separate dependent variables. Table 5–6 summarizes the separate ANOVAs performed on the rating variables. Of the 10 ANOVAs, 6 are statistically significant at the conventional .05 level. It is appropriate to perform multiple comparisons (t tests) of means for the four groups for these 6 dependent variables. Since our interest centers on superficial ordering of the groups, we shall note the rank ordering of the means only.

Table 5–7 displays averages for the four treatment groups on each of the dependent variables. Restricting comment to the six significant variables, the four groups rank order the same on mathematics, statistics, and science; educational psychology rates the most favorable, followed by commerce, speech, and criminal justice. An opposite ordering of means is seen for football rating. These four dependent variables compose the first multivariate dimension, which was labeled quantitative. The single significant dependent variable, included in the humanistic factor, was English literature. Groups are rank ordered from speech, with the highest mean to educational psychology, followed by commerce and criminal justice.

Table 5–6. Summary of Univariate Analyses of Variance on the 10 Dependent Variables

Variable	Treatment MS	Error MS	F
Mathematics	10.17	1.08	9.38***
Statistics	13.29	.70	18.91***
English literature	5.75	1.10	5.21**
Football	3.40	.96	3.53*
Basketball	2.30	1.23	1.86
Science	8.88	.85	10.49***
Rock music	.43	1.16	.37
Classic music	1.91	.98	1.94
Country music	2.36	1.46	1.62
Marijuana	4.53	1.26	3.61*

*$p < .05$. **$p < .01$. ***$p < .001$.

Table 5–7. Averages of the Four Treatment Groups on the 10 Dependent Variables

	Group			
Variable	Educational psychology	Speech	Criminal justice	Commerce
Mathematics	3.40	2.88	2.53	3.18
Statistics	3.98	3.25	3.02	3.53
English literature	3.52	4.38	3.23	3.29
Football	3.78	4.25	4.25	4.00
Basketball	3.72	3.81	4.13	3.76
Science	4.00	3.38	3.22	3.59
Rock music	3.69	3.75	3.65	3.41
Classic music	3.70	3.88	3.37	3.71
Country music	2.85	2.75	3.22	2.71
Marijuana	3.23	3.00	2.78	2.47

Although the third synthetic dimension does not differentiate the groups significantly, one dependent variable that defined the factor does do so. This variable, marijuana, rank orders the groups from educational psychology (the most favorable, although the average rating was 3.23—essentially neutral on the rating scale), the highest, to speech, followed by criminal justice and commerce.

Discussion The research was essentially exploratory in nature to determine if there might be significant separation between the groups on the miscellaneous attitude ratings. The statistically significant MANOVA and subsequent ANOVA tests indicate that chance alone would rarely produce the degree of separation found between groups on the 10 variables. However, since no hypotheses were formally stated for testing, the results of the study might best be regarded as offering a basis for more refined hypotheses to be tested in subsequent research.

The three synthetic variables that emerged from the MANOVA analysis were not totally unexpected. Over a period of years during which the same type of data has been collected from our statistics students, several factor-analytic studies based on all students have uniformly revealed the same three types of factors that were found in this study. However, for factors uncovered in a conventional factor-analytic design to emerge in a

MANOVA design, it is essential that the factor variables discriminate between the treatment groups. What is factored in MANOVA is the between-groups effect (treatment) relative to error dispersions. Since there was no a priori information to suggest that the three factor dimensions would discriminate between the four groups of majors, no such predictions regarding factor structure of the MANOVA were made. It should be noted that the structure of the research study was adequate to extract *three* factors. The number of factors that can be extracted is limited to the number of degrees of freedom for the treatment or the number of dependent variables, whichever is fewer.

On a purely post hoc basis, the synthetic variables make sense and represent an economy over the univariate consideration of each of the 10 dependent variables separately. The findings of the two approaches are consistent, but such is not always the case. For example, a variable may appear unimportant because it fails to separate the treatment groups. However, the same variable may correlate strongly with a variable that does separate the treatment groups and may therefore serve as a suppressor variable. The apparently unimportant variable suppresses error variance from the effective variable, enabling the latter to separate the treatment groups more effectively. We are able to separate the four groups significantly on a quantitative dimension and independently on a humanistic dimension. It seems intuitively compelling to view educational psychology majors as more theoretically (quantitative) inclined than the other three groups: commerce second, speech third, and criminal justice last. Likewise, the second dimension (humanistic) seems appropriately to place speech highest; however, the remaining groups do not seem as convincingly placed.

A brief comment is in order regarding the strong indication of heterogeneity of error-dispersion matrices. Group 3 (criminal justice) was noted as the source of a relatively large determinant that has the effect of making for a more conservative MANOVA test. Since the MANOVA tests were uniformly significant, little importance attaches to this finding. Of course, Group 3 (criminal justice majors) may differ psychometrically from the other three groups in important and interesting patterns of measurement.

Summary Four groups of students, majoring in educational psychology, speech, criminal justice, and commerce, were compared on attitude ratings toward 10 miscellaneous topics. Initial MANOVA tests indicate that the four groups differ significantly on the set of 10 measures. Subsequent comparisons of the four groups on synthetic and then on univariate variables were consistent in ordering the four groups.

Two-Way Factorial Design

Problem The following research is reported in Timm (1975). An experiment was designed to compare the effects of three class meeting time
during the day and two different teaching methods on achievement in a
physics course.

Method One independent variable consisted of three class meeting times
during the day—morning, afternoon, and evening. The second independent variable involved two different methods of teaching physics—a traditional approach and a discovery method. The design is schematized in
Figure 5–1.

Presumably the 24 students who served as subjects for the experiment
were randomly assigned to the six cells of the two-way factorial design,
with the restriction of ($n = 4$) subjects per cell. Three achievement tests
(one each on mechanics, heat, and sound) were administered as the
dependent variables during final exams.

Teaching Methods

	Traditional	Discovery
Morning	$n=4$	$n=4$
Afternoon	$n=4$	$n=4$
Evening	$n=4$	$n=4$

Figure 5–1. Schematic of Two-Way Factorial Design

Results The test of homogeneity of dispersion matrices, Box's F test, ielded an F of 1.83, with degrees of freedom of 30 and 732. This F value is ignificant at less than the .01 level. The results of this test indicate ignificant heterogeneity of dispersion (error) matrices. Table 5–8 dis-lays the determinants of the error-source-dispersion (within-cells) natrices. It is apparent that the afternoon class, which was taught by the liscovery method, is represented by an extreme outlier dispersion matrix.

Table 5–8. Determinants of Within-Cells Error-Source Matrices

	Traditional	Discovery
Morning	41.56	2.08
Afternoon	.14	1959.26
Evening	41.57	1.12

A summary of the MANOVA is shown in Table 5–9. All four MANOVA tests are in agreement in showing lack of statistical significance at the .05 level for the interaction between time of day and teaching methods. Accordingly the main effects of the two independent variables become the focus of interest. All four MANOVA test criteria are again consistent in indicating statistical significance (at the .001 level) for the main effects of time of day and teaching method.

For time of day, two significant factors (roots) were extracted. Table 5–10 displays the statistical tests on the roots and indicates that the preponderance of trace was accounted for by the first root, showing that variations in the data are associated almost exclusively with one measured achievement dimension.

Table 5–11 presents correlations between the two synthetic variables and the original dependent variables. The first synthetic variable is composed almost exclusively of the dependent measure heat. The second synthetic variable is predominantly composed of the mechanics measure but with a portion of the heat measure not correlated with the first synthetic variable. Note that the measure on sound is poorly represented in either of the two synthetic variables.

The main effects of time of day may now be examined in terms of the two synthetic variables. On the first synthetic variable, which was identified as a predominantly heat measure, the morning group is superior to the other two groups followed by the afternoon group and then the evening group. Table 5–12 gives the centroids of the groups. The critical difference required for significance between group centroids was computed by multiplying the standard error of the centroids with the t value

Table 5–9. Summary Table of Two-Way Factorial MANOVA

Source	Λ	df_n	df_e	F	S	M	N	Roy's GCR	Pillai's trace	Lawley–Hotelling trace
Time of day (row)	.006	6	32	64.385**	2	.0	7	.968**	1.787**	34.455**
Teaching method (column)	.058	3	16	87.206**	1	.5	7	.942**	.942**	16.351**
Interaction (row × column)	.549	6	32	1.864	2	.0	7	.430	.467	.792

*$p < .05$. **$p < .01$. ***$p < .001$.

Table 5–10. Statistical Tests of Synthetic Variables (Row) Main Effects of Time of Day

Root	% Trace	χ^2	df
1	86.87	58.339***	4
2	13.13	29.058***	2

*$p < .05$. **$p < .01$. ***$p < .001$.

Table 5–11. Intercorrelations between Synthetic and Dependent Variables

Variable	Synthetic variable	
	I	II
Mechanics	−.019	.993
Heat	.808	.302
Sound	.031	.076

required for significance at the .01 level. The procedure used is similar to the computation of Scheffé's t, and is the most conservative of multiple-comparison tests. This particular procedure was selected because there was no a priori basis for assigning particular weights to variables for determining the synthetic scores and because no a priori grounds were given for rank ordering the group centroids. The critical difference between centroids was found to be 2.6, a value that is clearly exceeded between each of the three group centroids. Therefore, we may regard the group centroids as being each significantly different from the others.

On the second synthetic variable, which is regarded as essentially a mechanics measure, the afternoon group scored highest, followed by the evening and then the morning groups. Examination of the differences

Table 5–12. Group Centroids on the Two Synthetic Variables for Different Times of Day

Group	Synthetic variable	
	I	II
Morning	19.022	12.979
Afternoon	13.485	16.945
Evening	7.421	13.100

between group centroids in terms of the critical difference computed for the first dimension (CD = 2.600) indicates that the afternoon group scored significantly higher than either the morning or evening groups however, the morning and evening groups do not differ significantly.

An alternative strategy for examining the results is to focus on the separate univariate findings. In view of the significant MANOVA findings and in accordance with the Hummel–Sligo (1971) strategy, univariate F tests were performed on the dependent variables separately. The outcome of this effort is anticlimactic, in that the synthetic variables were so closely identified with separate dependent variables.

As shown in Table 5–13, the mechanics and heat tests significantly separated the three groups, whereas the sound measure failed to separate the groups significantly. For the mechanics measure, the critical difference between means (at the .01 level) computed via the Scheffé t is 5.146. As seen in Table 5–14, the afternoon group scored significantly higher than the morning and evening groups; however, the morning and evening groups do not differ significantly.

For the heat variable, the required critical difference between means is 10.444. The morning group scored highest, followed by the afternoon and then the evening groups. Each group mean is significantly different from

Table 5–13. Summary of Univariate Analyses of Variance on the Three Dependent Variables

Variable	Treatment MS	Error MS	F
Mechanics	354.667	8.806	40.278***
Heat	6515.168	36.278	179.591***
Sound	7.542	15.236	.495

*p < .05. **p < .01. ***p < .001.

Table 5–14. Means of Three Achievement Measures According to Time of Day

Variable	Time of day		
	Morning	Afternoon	Evening
Mechanics	39.500	51.500	40.500
Heat	137.000	117.250	80.750
Sound	32.750	33.250	31.375

Table 5–15. Intercorrelations between Synthetic and Dependent Variables for Teaching Methods

Univariate variable	Synthetic variable I
Mechanics	.970
Heat	.466
Sound	.094

he other two group means. Since the univariate F failed to be significant or the sound variable, no comparisons between means were performed.

Next the main effects of teaching method are examined, first in terms of he synthetic variable of MANOVA and then in terms of the separate univariate dependent variables. Only one synthetic variable was extracted, and therefore the four MANOVA test criteria are equivalent in providing exact probability values.

Table 5–15 indicates the nature of the synthetic variable; it correlates strongly with the mechanics measure and moderately with the heat measure. The sound measure correlates poorly ($r = .094$) with the synthetic variable. Therefore, the synthetic variable appears to be primarily a measure of mechanics achievement with a small component of heat measurement.

The most conservative estimate of critical difference between centroids, computed on the assumption of post hoc selection of weights for the synthetic variable and no a priori planned comparisons, is 1.727. As noted in Table 5–16, the difference between the centroids is approximately 7 units, which far exceeds the difference required at the .01 level. Furthermore, the advantage is clearly in favor of the discovery method.

Examination of the separate dependent variables in univariate fashion according to the Hummel–Sligo strategy, results in a statistically significant F ratio for the mechanics and heat measures but not for the sound measure. The discovery method is clearly better for the mechanics and heat measures. (See Tables 5–17 and 5–18.)

Table 5–16. Group Centroids on the Synthetic Variable for Two Teaching Methods

Teaching method	Synthetic variable I
Traditional	13.334
Discovery	20.337

Table 5–17. Summary of Univariate Analyses of Variance on the Three Dependent Variables

Variable	Treatment MS	Error MS	F
Mechanics	2440.167	8.806	277.117***
Heat	2320.668	36.278	63.969***
Sound	40.042	15.236	2.628

$*p < .05. **p < .01. ***p < .001.$

Table 5–18. Means of Dependent Variables for Teaching Methods

Variable	Teaching method	
	Traditional	Discovery
Mechanics	33.750	53.917
Heat	101.833	121.500
Sound	31.167	33.750

Discussion The finding of significant heterogeneity of dispersion (error) matrices and the presence of a very obvious outlier cell deserve further attention. The direction of the outlier cell is toward a higher value, and since the error term in MANOVA results from a pooling of the separate error sources, the resulting F ratios on Wilks's lambda are conservative estimates. One might expect that the outlier cell would produce interaction between the two treatment variables. Since this was not the case, the finding of significant main effects for time of day and teaching methods appears even more convincing.

A number of comments seem in order regarding the design of the research project. The use of three different achievement measures (mechanical, heat, and sound) suggests that these are three important and presumably relatively independent measures of course achievement. The research design is defective in not permitting the extraction of as many synthetic variables. For the teaching-methods variable, only one root could be extracted, so that the three dependent variables were forced into one synthetic variable. Here again, four teaching methods would be required to permit three roots to be extracted. Of course, one of the achievement measures (sound) failed to discriminate between the treatment groups and would have failed, in any case, to emerge as a synthetic dimension.

Another comment in regard to a weakness in the research design is appropriate. The comment applies whether the design is univariate or multivariate; however, in the case of a multivariate design, because of its greater complexity, the weakness is much more easily concealed. The essential issue relates to the nature of the observed treatment effects, whether they are intrinsic or extrinsic. The issue here pertains to adequacy of control of Types S, G, and R errors.

With respect to Type S error, it was assumed that subjects were randomly assigned to cells, thereby controlling for subject-type variables such as age, intelligence, attitude toward the subject matter, and so forth. If the assumption of random assignment of subjects is not correct, the outlier error matrix may reflect variant subject characteristics as compared with the remaining cells of the design. With respect to Type G error, there is little that can be said, in view of the lack of a significant interaction between time of day and teaching methods. However, we must caution that the particular teachers and the manner in which they were assigned to the treatment cells could produce serious distortion of the treatment effects. For example, instructors who must teach at a preferred or nonpreferred time of day and use a preferred or nonpreferred method of teaching could produce an effect of teacher reaction that would be confounded with the nominal treatments under study in the design.

As for Type R error, previous experience and expectancies on the part of students and teachers with respect to time of day for class and to method of teaching could seriously affect the particular findings of this study and could make successful replication with a different group of students difficult. A frequent source of difficulty with the time-of-day variable (randomization of students to treatments was assumed as a control for this variable) concerns the fact that student selectivity (and perhaps teacher selectivity) is operating; for example, the better students may be taking a difficult elective course in the afternoon or evening and may thus be unavailable for the afternoon class.

In view of the previous comments, it is suggested that the significant effects of time of day and teaching methods on course achievement found in this study may actually be due to other variables that were confounded with treatments. The same kind of cautions apply to the research design whether it is univariate or multivariate.

Summary An experiment presumably involving the effect of time of day and teaching methods on achievement measures in three areas of course content indicated statistically significant MANOVA main effects for time of day and teaching methods. Both the synthetic and univariate measures were essentially consistent in displaying treatment effects.

Assuming that the results of the research are valid, the use of

MANOVA has permitted the investigator to evaluate appropriately the effects of the two independent variables on three dependent variables simultaneously. Taken at face value, the findings indicate that afternoon is the best time of day for teaching mechanics; however, for best achievement in heat scores, morning hours are best, followed by afternoon and then evening. Also, the discovery method is best for teaching both mechanics and heat.

Treatments × Subjects Design

Problem A study by Postovsky (reported in Timm, 1975, p. 228) tests the effects of delay in oral practice as learning of a second language begins. No research hypotheses or psychometric hypotheses with respect to four achievement measures are reported by Timm.

Method Matched pairs of subjects were formed on the basis of age, education, former language training, intelligence, and language aptitude. Presumably random assignment to treatments was performed for each matched pair. One group served as an experimental group and the other as a control. The experimental group experienced a 4-week delay in oral practice, whereas the control group experienced no delay. Treatment effects were sought under four measurements of language skills: listening (L), speaking (S), reading (R), and writing (W). These measures were taken at the end of 6 weeks of study.

Results No test of homogeneity of dispersion matrices (error) was performed, since the error term derived from a pooling of Matched Pair × Treatment interaction for each pair of matched subjects.

Table 5–19 provides the results of the MANOVA tests of statistical significance. The *F* ratio on Wilks's lambda failed to reach significance at the .05 level. Since only two treatments were conducted (experimental vs. control), only one synthetic variable could be derived, and consequently the four MANOVA test criteria are equivalent. Therefore the null hypothesis, which asserts that the two groups do not differ on the four dependent variables, is retained for all MANOVA significance tests. Under these circumstances, it is not appropriate to examine the synthetic variable.

In view of nonsignificant MANOVA tests, and following the Hummel–Sligo (1971) procedure, it is inappropriate to examine the four separate dependent variables with univariate *F* tests.

Discussion Although the MANOVA tests indicate failure of the four dependent variables to separate the experimental group from the control

Table 5–19. Summary of Multivariate Analyses of Variance on the Four Dependent Variables

Source	Λ	df_h	df_e	F	S	M	N	Roy's GCR	Pillai's trace	Lawley–Hotelling trace
Oral practice	.738	4	24	2.133	1.0	1.0	11.0	.262	.262	.356

$*p < .05. **p < .01. ***p < .001.$

group significantly, it is of interest to view the univariate F tests performed on the separate dependent variables. Table 5–20 presents a summary of the four analyses of variance. F ratios associated with two of the dependent measures, speaking and writing, were statistically significant at the .05 level. In view of the nature of the experimental treatment (oral practice of the foreign language), it is not surprising that the control group, which began oral practice of the language earlier than the experimental group, surpassed the experimental group in speaking ability. The mean for the control group (shown in Table 5–21) was 48.643, and that for the experimental group was 44.929. Likewise, although it seems less intuitively clear, the control group surpassed the experimental group on writing skill; the control group averaged 86.679, and the experimental group averaged 81.071.

Aside from the inherent interest value of the univariate findings, there are three additional matters of interest. First, the analysis of the data for the matched pairs design is identical to a repeated-measures-type design in which the subject serves as his or her own control. Thus the sample problem provides a model for this type of design. Second, this problem

Table 5–20. Summary of Univariate Analyses of Variance of the Four Dependent Variables

Variable	Treatment MS	Error MS	F
Listening	5.786	19.564	.296
Speaking	193.143	31.736	6.086*
Reading	34.571	11.646	2.969
Writing	440.160	71.420	6.163*

*$p < .05$. **$p < .01$. ***$p < .001$.

Table 5–21. Means, Standard Deviations, and d Measures of Dependent Variables for Control and Experimental Groups

Variable	Group M Control	Experimental	SD	d
Listening	29.143	28.500	4.423	.145
Speaking	48.643	44.929	5.633	.659
Reading	35.893	34.321	3.413	.460
Writing	86.679	81.071	8.451	.663

serves as a useful example of the occurrence of significant univariate findings when the overall MANOVA findings are not significant. The proliferation of different dependent variables will virtually assure the investigator of finding *something* statistically significant. Such a procedure inflates the nominal alpha level that the investigator wishes to control. Insistence on a significant MANOVA test prior to examining separate univariate ANOVAs keeps the alpha level close to the nominal level.

A third important issue pertaining to failure to find statistical significance in the overall MANOVA test is the matter of statistical power. It is enlightening to determine the level of statistical power under which the research was presumably performed. If the statistical power of the research was small, perhaps replication of the research (taking care to increase the statistical power by one of the available means) should be considered. One such way of increasing power involves the use of a larger sample size determined by the calculations of the statistical power.

The rationale for the following method of estimating statistical power is outlined in chapter 3. Since the experiment involved only two treatments, the procedure for power estimation is very similar to that for ANOVA. A value for phi is calculated and is used to enter the appropriate Pearson–Hartley power chart, which is selected according to degrees of freedom for the treatment and error terms.

$$\phi = [\delta^2/(P + 1)]^{1/2}$$
$$\delta^2 = \text{Hotelling } T^2 = df_{\text{error}} \cdot \lambda_1$$
$$\lambda_1 = \text{eigenvalue resulting from factoring product matrix}$$
$$(SSCP_{\text{error}}^{-1} \cdot SSCP_{\text{treatment}})$$
$$df_{\text{error}} = (\text{number of matched pairs} - 1)$$
$$p = \text{number of dependent variables}$$
$$\delta^2 = 27 \,(.3555) = 9.5985$$
$$\phi = [9.5985/(4 + 1)]^{1/2} = 1.3855$$

The Pearson–Hartley power chart for ($df_1 = 4$) and ($df_2 = 27$) was entered using ($\phi = 1.3855$). Statistical power of the experiment is estimated as approximately .62. Assuming that the population treatment matrix and pooled error-dispersion matrix are the same as in the research sample, the probability of detecting the real difference is .62; or conversely, the probability of failing to detect the real difference is .38. We find then that although the Type I (or alpha) error is set at a conventional small value (.05), the Type II (or beta) error is much larger (.38).

By inserting trial values of degrees of freedom into the formula for delta-square, it is possible to estimate how large a sample is required to detect real differences, assuming the statistical power is set at .95 (or Type II error is controlled at .05). A sample size sufficient to detect real differences with .95 power is found to be approximately 62 subjects who serve as their own controls, or 62 + 62 matched subjects = 124 subjects. Whether a replication of the experiment with about 124 subjects is worth pursuing depends on many considerations. One such consideration is the relative size of the treatment mean differences, that is, the effect size relative to the estimated standard deviation $[(M_1 - M_2)/SD]$. Cohen (1977) suggests that effect sizes be graded into three categories: small (.2 or .3 times standard deviation), medium (.5SD), or large (.8SD). Using this approach, the effect sizes for the separate dependent variables range from small for the listening measurement to medium for speaking, reading, and writing measures. This finding suggests that a replication of the study might be worthwhile.

Summary An experiment focused on the effects of delay in oral practice during learning of a second language. A control group experienced no delay in oral practice, whereas an experimental group experienced a 4-week delay. Four language skills were measured after 6 weeks of language training. MANOVA tests were not significant, and the univariate tests were deemed inappropriate. Statistical power of the MANOVA test was calculated, and an estimate of sample size needed for .95 power was determined.

Mixed Design: One Between-Subjects and One Within-Subjects (Lindquist Type I Design)

Problem An experiment cited by Timm (1975, p. 490) compares two new methods of teaching reading and mathematics with an older conventional method of teaching. Students were measured twice—once before treatment and once after. Neither experimental nor psychometric hypotheses were stated.

Method Students in a fourth-grade class were randomly assigned to one of three treatment groups with the restriction that there be $(n = 10)$ students per group. One group of students was taught by the customary, conventional method; the other two groups were taught by two different methods.

Two achievement tests, one in reading and the other in mathematics, were administered to obtain grade-equivalent levels for the students at the beginning of instruction and again at the end of instruction, 6 months later. The grade-equivalent scores, which were carried to one decimal place, were used as measures of the dependent variables.

Table 5–22. Tests of Homogeneity of Dispersion Matrices and Determinants Associated with the Two Error Sources

	Error source	
	Between subjects	Within subjects
Determinant		
1	.108	.004
2	.050	.002
3	.015	.005
Box test		
F	1.192	1.014
df_n	6	6
df_d	18168	18168

$*p < .05. **p < .01. ***p < .001.$

Results The design consists of two parts, a between- and a within-subjects analysis, each based on a different error term. Tests of homogeneity of dispersion matrices are carried out separately for each part.

The Box F test (Box, 1949) of homogeneity of dispersion matrices (error sources) was not significant at the .05 level for either the between- or within-subjects error matrices. Therefore, the hypothesis of homogeneity of dispersion matrices (error) is retained. Table 5–22 displays both the Box F tests and determinants for the error sources.

Table 5–23 presents results of the MANOVA tests of statistical significance. The interaction between test period (before or after) and treatment group is significant at less than the .01 level for each of the MANOVA tests. In view of the significant interaction, interest is centered on the simple effects involving change from pretest to posttest means according to treatment group.

The statistical analysis of the interaction resulted in extraction of two synthetic variables, but only the first was significant at the .05 level. Table 5–24 displays the significance tests on the two roots.

The nature or name of the synthetic variable is now of interest. Table 5–25 displays the correlations between the synthetic and dependent variables. The correlations suggest that the first synthetic variable is essentially the reading measure, whereas the second synthetic variable is more allied with the mathematics variable after suppression of a sizable portion of the reading measure.

Interest centers on only the first synthetic variable. Figure 5–25 displays the group centroids for pretesting and posttesting times. It is evident that

Table 5–23. MANOVA Summary Table

Source	Λ	df_n	df_e	F	S	M	N	Roy's GCR	Pillai's trace	Lawley–Hotelling trace
Treatment group	.338	4	52	9.377**	2	-.5	12	.642**	.699**	1.855**
Test period	.059	2	26	207.965***	1	.0	12	.941**	.941**	15.997***
Treatment × Test period	.235	4	52	13.845**	2	-.5	12	.735**	.850**	2.903**

$*p < .05.$ $**p < .01.$ $***p < .001.$

Table 5–24. Statistical Tests of Synthetic Variables Derived from Interaction of Treatment Groups and Testing Times

Root	% Trace	χ^2	df
1	95.52	35.190**	3
2	4.48	3.241	1

$*p < .05.$ $**p < .01.$ $***p < .001.$

Table 5–25. Intercorrelations between Synthetic (Discriminant) and Dependent Variables

	Synthetic variable	
Variable	I	II
Reading	.910	−.414
Mathematics	.293	.956

all three groups improved, but the greatest gain was made by Group 3 (Method B), followed by Group 1 (conventional). Recognizing that no discriminant weights were posited for the synthetic variable and that no a priori hypotheses were declared for group outcomes, a conservative, post hoc critical difference between centroids was determined as 1.83 at the .01 alpha level. The difference between group centroids on pretesting and posttesting exceeded the critical difference for each of the three treatment groups. Thus a significant gain was found regardless of teaching method.

On the pretest the two experimental groups (Groups 2 and 3) were similar to each other but superior to Group 1. Since Group 1 was taught in conventional fashion, it may be regarded as a control group. The gain from pretest to posttest achieved by Group 1 can be used as a standard against which to evaluate gains of the other two groups. Using this procedure, only the gains of the (experimental) Group 3 over the gain of (control) Group 1 exceeds the critical difference.

According to the Hummel–Sligo (1971) strategy, it is now permissible, due to the significant MANOVA tests, to examine the dependent variables in univariate fashion. Table 5–26 displays the results of applying univariate F tests to each of the dependent variables. Since the two synthetic variables were clearly identified with reading and mathematics, respectively, it is not surprising to find that both reading and mathematics measures separated the three groups significantly.

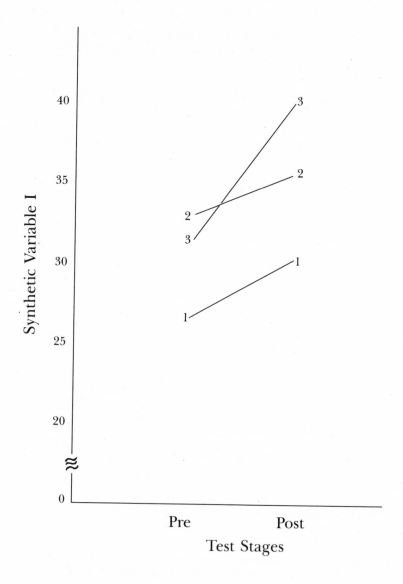

Figure 5–2. Plots of Centroids on Synthetic Variable I for Three Treatment Groups

Table 5–26. Summary of Univariate Analyses of Variance (Interaction of Treatment × Test Period) on the Two Dependent Variables

Variable	Treatment MS	Error MS	F
Reading	1.283	.041	31.323***
Mathematics	.561	.117	4.816*

$*p < .05. **p < .01. ***p < .001.$

Figures 5–3a and 5–3b depict cell means for pretests and posttests for each of the three groups for the two dependent variables. The general pattern of results for the reading variable closely parallels that for the group centroids. Mathematics, where both experimental groups make gains greater than that of the control group, is a slightly different pattern from the centroids. The critical difference for the Scheffé test at the .01 level is .300 for the reading measure and .506 for the mathematics score. Comparison of gain scores in reading and math of the conventionally taught Group 1 with gains of Groups 2 and 3 resulted in the following: On reading, Group 3 was significantly superior to Group 1, whereas Group 1 was superior to Group 2. On math, Group 3 was significantly superior to Group 1, whereas the gains of the control group and Group 2 did not differ significantly.

Discussion MANOVA tests indicated a significant interaction between testing (before or after) and the three treatment groups on the two dependent variables. On the MANOVA synthetic variable, all three groups made significant gains from pretest to posttest. However, the gain made by Group 3 was significantly greater than the gain made by the control, Group 1, and the control group's gain was greater than that of Group 2. This result suggests that the teaching method used for Group 3 is superior to the control condition and should be considered as a possible replacement for the conventional method. Of course, many issues must be considered in reaching such a decision, such as cost of implementing the new method as opposed to the conventional method and so forth.

One disturbing finding calls into question the preceding remarks. The experimental groups did not differ significantly from each other on the pretest but scored higher on the pretest than the control group. The difference may represent failure of randomization to match up the groups successfully and may thus indicate alpha error. On the other hand, the difference may be due to Type G influence of an unspecified nature

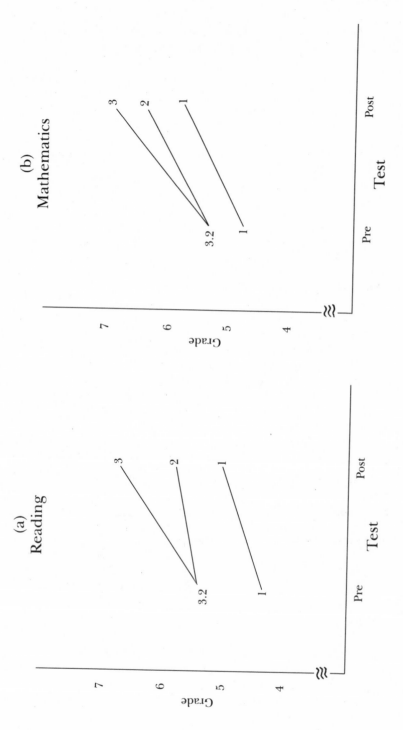

Figure 5–3. Plots of Cell Means for the Three Treatment Groups

Table 5–27. Tests of Homogeneity of Dispersion Matrices and Determinants Associated with the Two Error Sources

| | Error source | |
Cells	Between subjects	Within subjects
Deindividuation		
High	2.173	.027
Medium	.483	.016
Low	1.719	.017
Individuation		
High	14.123	.200
Medium	.969	.030
Low	5.846	.070
Box Test		
F	1.809*	3.670***
df_n	15	15
df_d	15949	143547

$*p < .05. **p < .01. ***p < .001.$

operative at the time of testing. Apart from the explanation of why the groups differed on the pretest measure, other important sources of Type G errors must be considered. For example, the instructors' attitudes toward the assigned or voluntary teaching methods could have been a potent source of influence on the teacher's teaching effectiveness and the students' posttest performance. As an example, suppose the three classes were taught at different times of day; the activity that preceded each of the three classes under study, such as scheduled recreation period as opposed to history instruction, could have seriously moderated the results.

An important phenomenon that often invalidates school research may have operated in this experiment: the spread of the treatment intended for only one group to other treatment or control groups. For example, children who are in one group may interact with children in other groups and may even provide practice in the different experimental methods under study. This kind of incident is particularly likely when two or more children from one family or neighborhood are assigned to different treatment groups. To prevent a diffusion of treatment effects between treatment groups, some variant of the groups' within-treatment design could be used.

One final class of cautions relates to attempts to generalize the research findings from one school to other schools and settings. Although randomization of students to treatment groups was carried out, no mention is made with respect to the nature of the population from which the students in the experiment were drawn. The class of concerns here expressed relate to Type R errors. The most effective teaching method found in this setting may be the least successful in a different setting with students of different socioeconomic levels and so forth.

Summary The effectiveness of two new methods of teaching fourth-grade reading and mathematics was compared with a conventional method in terms of standardized tests in reading and mathematics. Preliminary MANOVA tests indicated significant interaction between the three groups and pretest–posttest scores. Subsequent comparisons between the pretest and posttest scores for the three treatment groups, in terms of both a MANOVA synthetic variable and the separate reading and mathematics measures, were consistent in showing the superiority of the teaching method used with Group 3 over the conventional teaching method used with Group 1. Several cautions were raised regarding the validity of the results and possible generalization of findings.

Mixed Design: Two Between-Subjects and One Within-Subjects (Lindquist Type III Design)

Problem An experiment by Prentice-Dunn and Rogers (1980) tested one aspect of Zimbardo's deindividuation theory (1969). In accord with Zimbardo's theory, it was hypothesized that student experimenters would be aggressive toward a student subject when the experimenters were assured of anonymity and were relieved of responsibility for the aggression (deindividuation). In contrast, student experimenters who were required to assume responsibility for their aggression and were required to meet their subjects after the experiment were expected to be less aggressive than the former group (individuation). No specific psychometric hypotheses were stated with respect to two recorded outcome measures—shock intensity and shock duration. Actually, the instruction given to subjects specified the available levels of shock intensities that the subject could choose to administer and implied that the duration of all shocks would be constant at 1 second.

Method The research design is shown in Figure 5–4. Sixty male subjects were randomly assigned to the six treatment cells, with the restriction that ($n = 10$) subjects were retained for each cell.

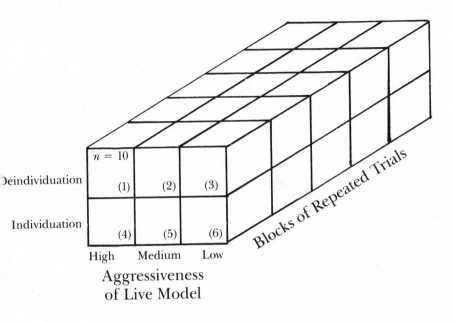

Figure 5–4. Mixed Design: A Two-Way Factorial with Repeated Measurement (Lindquist Type III). Numbers in parentheses identify cells.

The design is essentially a two-way factorial with repeated measurement on each subject. The two independent (factorial) variables consisted of type of experimenter input and the degree of aggressiveness displayed by a live model. The experimenter input involved two conditions: (a) deindividuation, which consisted of a gathering of variables thought to produce deindividuation, such as assurance of the student experimenter's anonymity and further assurance that the student experimenter would not be held responsible for his actions, and (b) individuation, such as the assertion that the student experimenter would be held responsible for his actions and that he would have to meet the student subject following the experiment.

The purpose of the experiment was falsified for subjects: They were told that they were assisting a biofeedback subject in maintaining his heartbeat rate at a predesignated high level. Whenever the heart rate fell below a predetermined level, the subject was to administer electric shock to raise the heartbeat rate. Actually, no shock was administered, and the subject was inaccurately informed that studies had shown that any shock intensity was adequate for raising the heartbeat rate.

The second independent variable involved a live model, who faked one of two levels of aggressive behavior toward the student subject: high aggressiveness, which involved selection of the most severe shock intensities during demonstration trials, and low aggressiveness, which involved selection of the least intense shocks for demonstration. A third control condition was added in which no live model was present. Following 5 demonstration trials by the live model for appropriate groups, student subjects performed independently for 20 trials. The trials were subsequently merged into four blocks of 5 trials each, the mean of each 5 consecutive trials to be used as a repeated measure for analytic purposes. Shock intensity and duration were recorded as two separate, concomitant dependent variables. However, the experimenters did not specify shock duration as a separate dependent variable. The measure was simply taken as an adjunct to measured shock intensity.

Results Since the design consists of a between- and a within-subjects component, the test of homogeneity of dispersion matrices must be carried out separately for each component. Table 5–27 presents the Box test for each component, together with the determinants of the error sources. For both the between- and within-subjects components, the Box test is significant at less than the .05 level, thereby indicating that the determinants differ significantly in both components. Examination of the determinant for both between- and within-subjects designs reveals that the determinant for cell 4 ("Individuation High") is clearly an outlier in both instances. Cell 4 involves the individuation treatment accompanied by a live model who displayed a high level of aggression toward the student subject.

In MANOVA, as in univariate ANOVA, it is customary to examine first the highest-order interaction; if significant, the simple cell means at this level are subjected to scrutiny and possible t tests. If the highest-order interaction is not significant, the next lower level of interaction is examined, and if it is significant, simple effects at this level are explored. Assuming that there are no significant interactions, main effects of variables are examined.

Following this procedure, it is noted in Table 5–28 that none of the MANOVA test criteria are significant for the triple interaction. The two-way interaction of Individuation × Trials was significant at the .05 level for three of the MANOVA criteria, Wilks's lambda, Lawley–Hotelling trace, and Pillai's trace. Only Roy's GCR failed to show significance at the .05 level. The reason for the failure of Roy's GCR to detect the difference is

Table 5–28. MANOVA Summary Table

Source	Λ	df_n	df_e	F	Roy's GCR	Pillai's trace	Lawley–Hotelling	S	M	N
Individuation (R)	.649	2	53	14.331**	.351**	.351**	.541**	1	0	25.5
Model (C)	.726	4	106	4.598*	.269***	.276***	.374**	2	−.5	25.5
Individuation × Model (R × C)	.939	4	106	.844	.061	.061	.065	2	−.5	25.5
Trials (S)	.814	6	322	5.819**	.183**	.187**	.228**	2	0	79.5
Individuation × Trials (R × S)	.922	6	322	2.212*	.053	.079*	.083*	2	0	79.5
Model × Trials (C × S)	.921	12	322	1.121	.056	.080	.084	2	1.5	79.5
Individuation × Models × Trials (R × C × S)	.915	12	322	1.214	.053	.086	.091	2	1.5	79.5

$*p < .05. **p < .01. ***p < .001.$

that the trace is not concentrated relatively in the first extracted root; the first root (eigenvalue) contained approximately 67% of the trace, whereas the second root contained approximately 33%. Although the main effects of model, individuation, and trials are significant by MANOVA tests, two of the variables (individuation and trials) are contained in the interaction just mentioned and are therefore ignored as main effects.

The only main effect significant by all MANOVA test criteria that is not contained within the significant interaction is the kind-of-model variable. We shall first examine fully the interaction effect between individuation and trials and shall then give full attention to the main effect of model.

The finding of a significant interaction between level of individuation and trials indicates that the pattern of trial means differs between the two groups, the deindividuated and the individuated. Although two roots were extracted, neither was significant at the .05 level ($\chi^2 = 8.757$, $df = 4$, $p = .067$ for the first root; $\chi^2 = 4.328$, $df = 2$, $p = .113$ for the second root). Note that the chi-square test of the first root is here equivalent to Roy's GCR test, both being based on only one root and both indicating lack of significance at the .05 level.

Since no a priori hypotheses were formulated regarding the nature of the two dependent variables, and consequently no commitment was given for a particular MANOVA test, it seems appropriate to take the three significant MANOVA tests seriously and to try to ascertain the nature of the significant interaction on the two synthetic variables. The use of both synthetic variables is made necessary by the lack of significant separation on either synthetic variable alone. Figure 5–5 depicts the group centroids (for individuated and deindividuated groups) on both synthetic variables across the four blocks of trials. It is apparent that the blocks of trials for the two groups form separate clusters that separate Group 1 from Group 2 on Synthetic Variable 2. However, Group 1 seems to comprise two subclusters: one cluster consisting of Trial Blocks 1 and 2, and the other of Trial Blocks 3 and 4. From these two subclusters it is evident that from the first two blocks to the last two blocks of trials, Group 1 increased on Synthetic Variable I while decreasing on Synthetic Variable II.

Group 2 also contains a subcluster consisting of Trial Blocks 1, 2 and 3. Trial Block 4 is an isolate. In contrast to the progression across trial blocks in Group 1, Group 2 showed no change in Synthetic Variable I but decreased on Synthetic Variable II across trial blocks. In substance, then, the significant interaction detected by the three MANOVA tests appears to result from the following: (a) for the individuated group (1), a simultaneous increase in Synthetic Variable I accompanying a decrease in Synthetic Variable II across blocks of trials, and (b) for the deindividuated group (2), no change in Synthetic Variable I, accompanying a decrease in Synthetic Variable II across blocks of trials.

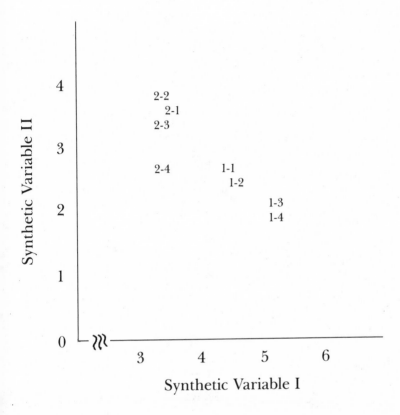

Figure 5–5. Centroids of Group 1 (Deindividuated) and Group 2 (Individuated) across Four Blocks of Trials. Labels 1–1, 1–2, 1–3, and 1–4 represent centroids of Group 1 on Trial Blocks 1, 2, 3, and 4, respectively.

Table 5–29. Intercorrelations between Synthetic (Discriminant) and Dependent Variables

	Synthetic variable	
Variable	I	II
Shock intensity	.996	−.086
Shock duration	.093	.996

From Table 5–29, which shows the intercorrelation between dependent variables and the synthetic variables, it is apparent that the first synthetic variable is virtually synonymous with the dependent variable, shock intensity ($r = .996$), whereas the second synthetic variable is just as impressively associated with the second dependent variable (shock duration; $r = .996$). The correlation of the two synthetic variables with the remaining dependent variable is in each instance virtually zero.

According to the Hummel–Sligo (1971) strategy, the significant MANOVA tests entitle the investigator to perform univariate F tests on the separate dependent variables. In view of the previously noted virtual identity between the two synthetic variables and the dependent variables, one would expect the univariate analyses to be anticlimatic. However, for the sake of illustration, the analyses will be treated in conventional fashion.

Table 5–30 presents a summary of the univariate F tests of the interaction between Individuation × Trial Block interaction. Only the first dependent variable, shock intensity, was statistically significant at the .05 level. Figure 5–6 depicts the cell means for shock intensity for the two levels of individuation across the four blocks of trials. The same pattern of changes across trials is noted for the two groups that was noted on the first synthetic variable; Group 1 increased shock intensity from the first two

Table 5–30. Summary of Univariate Analyses of Variance of Individuation × Trial Blocks

Variable	Treatment MS	Error MS	F
Shock intensity	5.455	1.826	2.988*
Shock duration	.044	.030	1.476

$*p < .05.\ **p < .01.\ ***p < .001.$

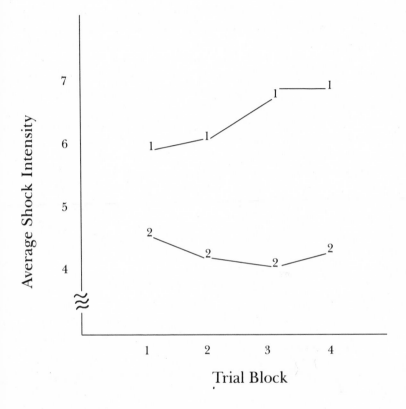

Figure 5–6. Mean Shock Intensity for Group 1 (Deindividuated) and Group 2 (Individuated) across Four Blocks of Trials

trials to the last two, whereas Group 2 maintained approximately the same intensity. The second dependent variable, shock duration, was not statistically significant and therefore, is not presented; however, the pattern of the cell means resembled the second synthetic variable, as expected.

The main effects of the levels of aggression associated with the live model was statistically significant by all four MANOVA criteria. Although two roots were extracted, only the first was significant at the .05 level. For the first root, $\chi^2 = 16.734$, $df = 3$, $p < .01$; for the second root, $\chi^2 = .384$, $df = 1$, $p > .05$. The nature of the first synthetic variable may be inferred from Table 5–31. Both shock intensity and shock duration correlate moderately with the synthetic variable and thus suggest a single punitive dimension. Although Synthetic Variable II does not separate the groups significantly, it is interesting to note that the two dependent variables again correlate moderately with the synthetic variable, except that shock duration serves as a suppressor variable.

Table 5–31. Intercorrelations between Synthetic (Discriminant) and Dependent Variables

Variable	Synthetic variable	
	I	II
Shock intensity	.772	.648
Shock duration	.635	−.773

The group centroids on the first synthetic variable were as follows: high-aggression model = 2.196, no-aggression model = 1.756, and low-aggression model = 1.499. The most conservative critical difference was computed as .611. With this criterion, performance of the group in the high-aggression model is significantly higher than that of the group with the low-aggression model; however, the no-aggression model, which is intermediate between the other two groups on the synthetic variable, does not differ significantly from the other two conditions.

Following a significant MANOVA test, the Hummel–Sligo (1971) strategy permits separate univariate analysis of the two dependent variables. Table 5–32 shows that each of the two dependent variables separated the three model conditions signficantly at less than the .02 level. Employing the Scheffé t, the critical difference required for significance at the .05 level was 1.319 for shock intensity and .238 for shock duration. The groups associated with high- and low-aggressive models differed significantly on both shock duration and shock intensity (also see Table 5–33). However, the control group did not differ significantly from the groups with models

Table 5–32. Summary of Univariate Analyses of Variance of Main Effects of Live Models

Variable	Treatment MS	Error MS	F
Shock intensity	66.525	11.097	5.995**
Shock duration	1.539	.360	4.281*

*$p < .05$. **$p < .01$. ***$p < .001$.

Table 5–33. Means of Dependent Variables for Main Effects of Live Models

	Aggression levels of live models		
Variable	High	None	Low
Shock intensity	6.330	5.360	4.508
Shock duration	.707	.501	.443

on either of the dependent variables. Use of a Scheffé-type test for both the multivariate and univariate comparisons seems desirable in view of its manifest robustness when error sources lack homogeneity.

Discussion There is need to examine carefully the extent to which the research met the basic assumptions of MANOVA. The Box F test, performed on between- and within-subjects error sources, indicates violation of the assumption of homogeneity of dispersion matrices. Typically MANOVA appears to be robust under violation of its assumptions (Ito & Schull, 1964), especially when an equal number of subjects are contained in each cell and there is no pattern to the violations. In this instance, the source of the heterogeneity for both between- and within-subjects was traced to an outlier cell 4, which produced an extremely large determinant in each instance. Such a condition produces conservative tests of statistical significance for both the between- and within-subjects effects; furthermore, an outlier cell may be of great interest, especially if treatment effects could possibly have produced it. For this reason, it is desirable to explore further the structure of the variance–covariance matrices.

Table 5–34 contains the variance–covariance matrices from the error sources. The outlier group of interest (cell 4), which received individuation influence and was exposed to a highly aggressive live model (within-subjects error source), is seen to be similar to other within-subjects error sources on the covariance of Dependent Variables 1 and 2. These covariances range from a high of .033 to a low of .003. Such low covariances indicate, as far as within-subjects variability is concerned, that there is essentially no correlation between the measures of shock duration and shock intensity within any of the error sources. However, inconsistency within subjects' performance, reflected by the variances for Dependent Variables 1 and 2, is significantly larger in the outlier group than in the other error sources.

The F_{max} test was used to evaluate the differences in group variances; for shock intensity $F_{max} = 3.76$, $df = 27$, $p < .05$; for shock duration $F_{max} = 8.524$, $df = 27$, $p < .01$. It appears that the individuating condition, coupled with the presence of a highly aggressive model, produced a higher level of variability (subject inconsistency) in the outlier group than in the remaining error sources. This suggests a conflict on the part of the subjects in the outlier group as to which set of circumstances was more compelling—taking responsibility for administering punishment or the need to perform in agreement with (imitate) the live model.

An examination of the between-subjects error sources of Table 5–34 discloses an interesting contrast between the outlier cell discussed (cell 4) and the group that received individuating influences and was exposed to a live model who performed at a low-aggression level. It should be noted that the data presently under discussion result from variability (error) between subjects within treatment cells after the repeated measures for each subject have been totaled into a single measure. In the outlier group, variance for shock duration remains significantly larger than any other cell variance—$F_{max} = 15.641$, $df = 9$, $p < .01$. Variances for shock intensity are made homogeneous by totaling blocks of trials into a single measure.

Although covariance between shock intensity and shock duration varies markedly between the outlier group (covariance is $-.891$) and for the individuated group with a low-aggression model (covariance is .832), a statistical test of the resulting correlations was not significant at the .05 level. The latter statistical test is of necessity based on a small number of subjects. It is interesting to speculate on the meaning of such discrepant covariances, assuming they are real. Of particular interest is the opposite sign associated with the two covariances. Apparently the treatment influences in the outlier group resulted in a reduction in shock intensity but produced a compensating longer shock duration, whereas in the other group, the treatment produced an increase of both shock duration and shock intensity. These findings suggest that the student experimenters experienced conflict between their perceptions of the actions of a live model and the realization that the experimenter must assume responsibility for administering the punishment.

The inconsistent MANOVA findings on statistical significance for the

Table 5–34. Variance–Covariance Matrices Computed from Error Sources

	High		Medium		Low	
	Intensity	Duration	Intensity	Duration	Intensity	Duration
Error source (within subjects)						
Deindividuated						
Intensity	.732	.007	1.968	.003	1.482	−.008
Duration		.037		.008		.012
Individuated						
Intensity	2.856	−.010	1.954	.033	1.962	−.015
Duration		.070		.016		.036
Error source (between subjects)						
Deindividuated						
Intensity	4.945	.382	7.371	−.001	11.260	−.003
Duration		.469		.066		.153
Individuated						
Intensity	14.561	−.891	6.875	−.110	21.570	.832
Duration		1.024		.143		.303

Level of aggregation (live models)

Individuation Level × Trial Blocks interaction are of considerable interest. Wilks's lambda, Lawley–Hotelling trace, and Pillai's trace each indicated significance at less than the .05 level. In contrast, Roy's greatest characteristic root failed of significance at the .05 level. It is well known that the four classic MANOVA tests of statistical significance differ in sensitivity (power) according to distribution of the trace into roots. Since Roy's GCR test evaluates only the first root, whereas the other three MANOVA tests represent a value based on all the roots, it is apparent that a strong concentration of the trace into the first root would make Roy's test maximally sensitive while decreasing sensitivity of the remaining tests (Barker & Barker, 1979). Harris (1975) expressed preference for Roy's GCR test due to the kind of situation that arose in this experiment. Although three of the MANOVA tests indicate statistical significance, no single synthetic variable reliably separates the relevant experimental cells. The significant separation resides in a simultaneous consideration of the two synthetic variables. The pattern of responses of the deindividuated group differs initially from that of the individuated group and becomes more different in later trials. On each trial the deindividuated group used greater shock intensity and lower duration than the individuated group and then increased the intensity of shock even further while decreasing shock duration on later trials. In contrast, the individuated group steadily maintained a low shock intensity during all trials but decreased shock duration on the last block of trials.

It should be noted that Prentice-Dunn and Rogers did not choose to view the within-subjects component of the data. They were in possession of a greater number of subjects and collected additional data with respect to the subjective states that the subjects reported undergoing during the experiment.

One major purpose of the experiment was to examine how a live model who displayed different levels of aggression affected performance of subjects. This variable was found to be statistically significant on all four MANOVA tests and on the subsequent univariate F tests. The one significant synthetic variable extracted by MANOVA was related to both shock intensity and duration and therefore may be regarded as a simple noxious punishment variable. The most conservative critical difference between centroids required for significance at the .01 level was computed to be .610. By this standard, only the difference between centroids of the high- and low-aggression models attained significance. The group exposed to the high-aggression model administered most punishment, whereas the group exposed to the low-aggression model administered the least. The no-aggression model group was intermediate between the high- and low-aggression groups. Examination of the dependent variables in univariate fashion reveals precisely the same pattern as for the synthetic variable.

However, on shock duration, difference between the three groups does not exceed the critical difference.

Perhaps the most interesting finding, with respect to the effect of a live model, is that the no-model group performed at an intermediate level between the high- and low-aggression model groups. This finding supports the view that the model exerts a potent effect in reducing aggressive performance when the model assumes a low-aggression stance. It is also of considerable interest that the effects of the live model did not interact significantly with the level of individuation.

Summary A Lindquist Type III design using blocks of repeated trials tested Zimbardo's theory of the effects of levels of deindividuation on aggression. Three levels of aggressive behavior of a live model constituted a second independent variable. Two measures of experimental outcome were shock intensity and shock duration.

Significant MANOVA effects were found for the levels of Deindividuation × Trial effect and for the main effect of live model. In general, it was found that conditions that produce deindividuation produce higher levels of aggression, but in terms of the pattern of shock intensity and duration, specific changes take place as trials continue. The effects of the live model did not interact with other experimental conditions but produced effects that were interpreted as indicating imitation on the part of the group.

6

Application of MANOVA to Univariate Designs That Involve Repeated Measures

Statisticians have long expressed reservations about the use of ANOVA designs that involve repeated measurements of the same dependent variable. This chapter will define the basic problems involved in such designs and will examine alternative univariate and multivariate approaches for solving them. Then we will explain the procedure for applying MANOVA to this type of univariate design and will describe the resulting advantages.

Distinction between MANOVA Applied to Univariate and Multivariate Repeated-Measures Designs

A repeated-measures design involving more than one dependent variable, like other designs covered in the text, can be analyzed in MANOVA fashion. Three examples are given in the previous chapter: the Treatments × Subjects design and two mixed designs, Lindquist Type I and Lindquist Type III. This chapter focuses attention on the basic unit of such designs: a strictly univariate design, wherein each subject serves as his/her own control and is measured repeatedly on the same dependent variable. It is important to note that the material of the present chapter applies also to a design of the matched pairs type.

Univariate Analysis of Repeated Measures

The simplest case of this design is the t test for the subject as his or her own control. Its formula

$$t = \frac{M_1 - M_2}{\sqrt{S_{M_1}^2 + S_{M_2}^2 - 2r_{12}\,S_{M_1}\,S_{M_2}}}$$

provides for the two treatment effects that are being compared and also for a correlated component, which is subtracted from the error term, possibly increasing the statistical power of the test. An important feature is that there is only one correlation involved between the pair of scores obtained for each individual.

Extension of the t test for the subject as his or her own control to the Treatments × Subjects design simply involves taking additional measures of the dependent variable for each individual. Although the additional measures are usually associated with different treatment conditions, repeated measures may be taken under one and the same treatment condition in order to measure cumulative effects, as in learning research. The Treatments × Subjects design, by introducing more than two repeated measures, produces a matrix of correlations between all possible pairings of repeated measures. For example, a Treatments × Subjects design consisting of three repeated measures produces a (3 × 3) correlation matrix between repeated measures. Assuming that a preliminary F test indicates that a t test is appropriate, a t test for the subject as his or her own control that compares means of Repeated Measures 1 and 2, 2 and 3, and 1 and 3 may show very different degrees of correlation. This range in correlation poses a logical difficulty in interpreting the three comparisons. Also, it is apparent that the preliminary test (analysis of variance) must have utilized some sort of average of the three intercorrelations.

The crux of the problem may be phrased as a question: How well does the average correlation represent individual intercorrelations of the matrix? This issue, as formulated by the statistician, is termed the assumption of homogeneity of covariance. Unless the intercorrelations between repeated measures can be adequately represented by some "typical" measure, the general Treatments × Subjects design should not be used. Thus the analysis of repeated-measures designs assumes not only the usual homogeneity of error variance but also the homogeneity of covariance of error sources. These two assumptions are termed *compound symmetry* when applied to the error matrix of variances and covariances. The necessity for the latter assumption and the difficulty in testing it have been responsible for the principal misgivings of statisticians about the repeated-measures designs.

Extension of the Treatments × Subjects design to higher-order designs requires, in addition to homogeneity of variance and covariance of error sources, the homogeneity of the separate error sources to be pooled to an average error matrix. Tests by Bartlett and Box (Box, 1949) were designed to accomplish these purposes.

A Univariate Procedure for Analyzing Repeated-Measures Designs

Huynh and Feldt (1970) demonstrated a simple method of analyzing repeated-measures designs that protects against lack of homogeneity of variance–covariance matrices. The method is applicable to any repeated-measures design. Two extremes of a quantity ($\hat{\epsilon}$) are considered; one extreme takes the value 1.00, and the other extreme is computed as a fraction: $\hat{\epsilon} = 1.00/(\text{number of repeated measures} - 1.00)$. The two values of $\hat{\epsilon}$ are used separately. Under the first value of $\hat{\epsilon} = 1.00$, homogeneity of variance–covariance matrix is assumed (i.e., multiplication of the $\hat{\epsilon}$ value by the degrees of freedom for treatment and then separately for error results in no change in the degrees of freedom). The F table is entered with the conventional degrees of freedom, and a standard interpretation is made.

Under the second value of $\hat{\epsilon}$, the worst possible violation of the assumption of homogeneity of the variance–covariance matrix is assumed; accordingly, the degrees of freedom for the treatment and error term are reduced to provide an appropriately conservative F test. The downward adjustment of the degrees of freedom is accomplished by multiplying the $\hat{\epsilon}$ value first by the treatment degrees of freedom, then separately by the error degrees of freedom. Essentially, these calculations produce degrees of freedom that are appropriate for a reduced design consisting of a t test for the subject as his or her own control, which requires no assumption of homogeneity of the variance–covariance matrix. The F table is then entered with these reduced degrees of freedom to provide a conservative test.

One of three important outcomes may result from application of the Huynh and Feldt approach. The first outcome involves failure to find significance with the conventional degrees of freedom ($\hat{\epsilon} = 1.00$). This condition assumes that the variance and covariance matrices are homogeneous. If so, there is no point in performing the more conservative test. In the second outcome, significance is found under the conventional F test but not under the more conservative test. Under these circumstances, the investigator may conclude that the probability lies somewhere between the chosen alpha level and that of the conservative test.

A method exists (Collier et al., 1967) for calculating an estimate of $\hat{\epsilon}$ that yields adjusted degrees of freedom such that the alpha level is reasonably approximated. The updated version of the computer package BIOMED (BMDP2V; Dixon, 1979) provides the vaue of $\hat{\epsilon}$ for repeated-measures designs.

In the third outcome, statistical significance may be obtained under both the conventional F test and the conservative F test. The investigator may then proceed with assurance that treatment effects were demonstrated at the chosen alpha level despite any possible failure of the homogeneity-of-variance–covariance assumption.

The Huynh and Feldt procedure provides a simple, rough rule of thumb for dealing with repeated-measures designs. Its simplicity results from avoiding a direct confrontation with the variance–covariance matrix, a practice that may result in a loss of valuable information regarding treatment effects. For this reason, a method that examines the variance–covariance issue directly will be considered next.

Multivariate Analysis of Variance of Repeated-Measures Designs

Two preliminary steps are required to prepare the repeated measures for MANOVA analysis. First, the differences between adjacent repeated measures are obtained for each individual, that is, the difference between Measures 1 and 2, between Measures 2 and 3, and so forth. The obtained difference scores are then substituted as different dependent variables in a MANOVA analysis. This procedure provides a test of the null hypothesis that treatment groups have parallel profiles on the repeated measures. It is possible to apply this MANOVA step only to higher-order repeated-measures designs.

A second step in applying MANOVA to the repeated-measures design involves using each repeated measure as a separate dependent variable. This step permits a test of the null hypothesis that treatment-group profiles of repeated measures are at the same level. It should be noted that the multivariate test is, in this instance, equivalent to the univariate test in which repeated measures are first totaled for each subject. The two steps, the first testing for interaction and the second for main effect, follow the traditional ANOVA sequence. It should be noted that tests of this kind are frequently referred to as constituting *profile analysis*.

As a consequence of applying MANOVA to the repeated-measures analysis, the dimensionality of the ANOVA design is reduced. For example, a Lindquist Type I (one between-groups effect and one within-groups effect) design is reduced to a simple randomized design, with first the difference scores and then the repeated measures, respectively, serving as dependent variables. A Lindquist III design (two between-groups effects and one within-groups effect) reduces to a two-way factorial with more than one dependent variable.

Two major advantages attend the use of MANOVA rather than ANOVA in analyzing repeated-measures designs. First, MANOVA makes no assumptions regarding the homogeneity of the variance–covariance matrix, and second, a possibly more reliable and meaningful synthetic variable is available for use. Two vital assumptions attend the use of MANOVA. One assumption that has its counterpart in ANOVA pertains to the homogeneity of dispersion matrices that are pooled to form the error matrix. If this assumption is not met, a transformation of the measurement scale may produce homogeneity; if not, it may be possible to reduce the design to a simpler level at which homogeneity may be attained. The second assumption is that each error source is of full rank. In other words, for each error source, there must be as many subjects as there are measures serving as dependent variables. This requirement may eliminate MANOVA as an alternative analysis in some repeated-measurement designs. In others, it may be possible to retain MANOVA as an alternative analysis by combining repeated measures into blocks of trials prior to undertaking the analysis.

As noted previously, the use of MANOVA on a univariate repeated-measures design involves two analyses—one applied to difference scores and the other to repeated measures. It is helpful to consider outcomes of the two analyses separately and in the order in which they are carried out.

1. *MANOVA applied to difference scores: results nonsignificant.* If the overall multivariate test when applied to difference scores (differences between adjacent trials) is nonsignificant, the interaction between the repeated measures and the treatment being tested is regarded as nonsignificant. This statement is equivalent to saying that the profiles of the treatment groups, with repeated measures used as elements, are essentially parallel. Therefore, it is appropriate to test for treatment main effect on the repeated measures. This may be either an ANOVA or MANOVA test, since under a tenable assumption of no interaction (parallel profiles) the two tests are equivalent. The MANOVA approach tests the significance of treatment-group separation on a synthetic variable, whereas the ANOVA approach tests (overall F) the separation of treatment groups on repeated measures that are summed and averaged.

2. *MANOVA applied to difference scores: results statistically significant.* Assuming the MANOVA test applied to difference scores is statistically significant, the treatment-group profiles are then regarded as nonparallel across the repeated measures. The choice of a next step in data analysis is complicated by the availability of several interrelated options.

The first, a univariate strategy applied to repeated measures, would be appropriate only if one significant root had been obtained in the MANOVA tests. The univariate strategy might consist of t tests (independent groups) comparing relevant treatment groups on each separate repeated measure or performing t tests (subject as his or her own control) between pairs of repeated measures for each treatment group separately. Either of these strategies is appropriate because the MANOVA test indicates homogeneity of the variance–covariance matrix (one factor in the matrix) and provides an overall statistical test of all difference scores.

If more than one significant root is obtained from the MANOVA test, routine application of univariate ANOVA procedures to repeated measures is not appropriate. Assuming the MANOVA roots are interpretable, univariate analyses may be applied to appropriate factor clusters of repeated measures. However, it may be more appropriate to employ a purely multivariate strategy.

A multivariate strategy may be appropriately utilized in the event of the extraction of one or more significant roots. Synthetic scores then become the focus of interest. Greater reliability normally attends the use of synthetic scores composed of several measures, and when they are appropriate, they are thus to be preferred to single measures.

Initially an attempt should be made to interpret the synthetic dimensions. For example, fatigue effects may be produced by one treatment and rejuvenating effects by another, as evident from the shapes of the performance curves. Then the treatment groups should be compared on each separate synthetic score, with care being taken to interpret the synthetic scores as change scores. A plot of treatment groups in the dimensions of the synthetic scores can be enlightening, making apparent that a treatment may affect more than one synthetic dimension.

7

Checklist for the Investigator Conducting MANOVA Research

This final chapter should be of value to the investigator in two ways. First, it provides a succinct summary of practical and vital issues discussed at greater length throughout the text. Second, it affords a practical checklist that the investigator can consult prior to, during, and following the conduct of MANOVA research. Topics on the checklist are arranged in the order in which they will normally be most needed in conducting MANOVA research.

Decision to Conduct a Study or Experiment

The decision as to whether to conduct an experiment or a study is sometimes made by the investigator and is sometimes dictated by the nature of the research and variables involved. An interesting blend of the two is seen in an analysis of previously collected study data as pilot or preliminary research for use in designing a full-scale experiment.

The investigator has numerous alternatives in dealing with pure study data. For example, if a large body of study data is available, the investigator may well draw at least two random samples and may attempt to replicate findings of the two studies. Successful replication adds stature and credibility to study research.

Variables that characterize organisms (organismic variables), such as height, weight, sex, and so forth, when used as "independent" variables in research pose problems in that they cannot be truly manipulated by the investigator. As a consequence, subjects cannot be randomly assigned across that particular independent variable dimension, resulting in failure to control for Type S variables. The net outcome is that differences found to be associated with variation in organismic variables, used as independent variables, are difficult to interpret; differences may be actually due to unknown variables that are confounded with the organismic variable.

Scientific research ranges from "pure" experiments to total studies but for the most part consists of intermediate combinations of the two. Research involving both study and experimental variables simultaneously in the same design requires greater caution in interpreting the study variables.

Selection of Dependent Variables

Contrary to usual practice, it is the contention of the authors that as much care should be exercised in selecting dependent measures as in designing the research. The qualities of good psychometric measures (reliability and validity) are essential to detecting treatment effects. Factor measures represent the epitome of measurement ideals and should be used where possible. If factor measures are not available, then pilot research in which good measures are identified and/or developed should be considered. Additional considerations arise depending on whether the investigator hypothesizes one or more dimensions of treatment effect.

1. *One dimension of treatment effect.* If previous research in the area of interest has employed a variety of measures designed to measure the same outcome variable, and the results have been consistent, the hypothesis of one dimension is supported. It would then be appropriate and wise for the investigator to use three or more measures (possessing best psychometric qualities) of the dimension of interest. Use of multiple outcome measures provides not only greater reliability for the one-dimensional synthetic variable but also an opportunity to verify empirically the unidimensionality of the set of dependent variables. The correlation matrix that results from pooling the *SSCP* matrices from error sources is the appropriate target for this factor analysis.

Multiple outcome measures provide a further opportunity to confirm that treatment effects are unidimensional across the treatment variables. The number and nature of significant synthetic variables resulting from the MANOVA procedure tests this hypothesis. Sufficient degrees of freedom associated with the treatment effect should be available to make this test possible.

2. *Two or more dimensions of treatment effect.* Two major instances require that the investigator evaluate two or more dimensions of measurement. First, extant theory predicts that treatment effect may differ along several measured dimensions (e.g., different measures of performance during latent learning, as predicted by Hullian theory). Second, accumulated research findings from similar research designs are inconsistent, even

though one and the same dimension of measurement is apparently employed for the dependent variable. To characterize this situation the same conceptual label is applied to the dependent-variable measures, but the operations for the measures may be quite different from one piece of research to another, and the reliability and validity of the resulting measures may vary widely. In this situation, it is conceivable that inconsistent findings may result from poor measures of one and the same dimension or from inadvertent measurement of more than one dimension, each of which is differentially affected by the treatments.

For each hypothesized dimension of measurement, the experimenter should attempt to provide at least three different measures. This minimum number will expedite the recognition and identification of a factorial dimension, both at the level of testing the dimensionality of the dependent variables and at the level of testing the dimensionality of treatment effects. Previously developed factor scales should be used when available and appropriate. The advantage of using a priori factor scores is so compelling that, in the absence of previously developed factor scales, the investigator should seriously consider conducting preliminary research in an effort to develop them.

Selection of a MANOVA Test Criterion

MANOVA test criteria are completely equivalent if only one dimension is extracted from the set of dependent variables; it does not matter which is used. Otherwise, the MANOVA test criteria differ, and the appropriate one should be chosen during the research design stage. It should provide a maximally sensitive (most powerful) statistical test of the hypothesized dimensionality of the dependent measures set. Three rough rules of thumb are offered for selecting the most appropriate MANOVA test.

1. Roy's GCR (theta). This test should be employed to confirm a hypothesis of one single dimension (or one predominant factor) in the dependent variable set.

2. Wilks's lambda. This test is maximally sensitive when two or more dimensions are contained in the set of dependent variables and are of relatively equal importance in accounting for the trace.

3. Lawley–Hotelling trace (tau) and Pillai's trace (V). These two test criteria appear to be intermediate in sensitivity when compared with Roy's GCR and Wilks's lambda. However, there is evidence that Pillai's trace criterion may be more robust to lack of homogeneity of dispersion matrices than the other three MANOVA criteria (Olson, 1976), but see also Stevens (1979).

Statement of Problem

The authors contend that hypotheses as to dimensionality of the set of dependent variables should be included alongside hypotheses regarding research outcomes. The two sets of hypotheses should be integrated in such a way that research hypotheses pertain to a particular dimension of outcome variables.

Research Design

The statistical design to be used for MANOVA research may be selected simply by ignoring the multivariate nature of the design and selecting an appropriate ANOVA design involving one dependent variable. This procedure is permissible because ANOVA and MANOVA are identical in underlying design structure.

With respect to treatment effects that are of primary interest, the investigator should plan enough treatment levels so that the dimensionality of the set of dependent measures can be statistically tested. A very similar practice in ANOVA involves providing a sufficient number of treatments to make possible a test of trend.

Computer Program Test

Although errors from hardware computer operation are extremely rare, ample opportunities exist in computer data processing for human error; these include mistakes in the recording and keypunching of data (perhaps the greatest error source), and computer program instruction errors. Fortunately, simple precautions that the investigator can take will provide a high level of protection from computer program (and computer) errors. The authors suggest that the investigator routinely obtain data sets that correspond to research designs used. These data sets may be obtained from statistics textbooks (or from hand-calculated examples). Both the computer program and the computer are tested by submitting to the computer the sample problem for the current research design and then evaluating the accuracy of solution. It is good practice to submit to the computer the sample problem again just prior to the current research data to assure that the example and the real data are processed at very nearly the same time.

Selection of MANOVA Strategy

The investigator should adopt a complete research strategy prior to examination of research data. The strategy should include the designation of a preliminary multivariate test and provision for appropriate follow-up synthetic and/or univariate comparisons. Strategies that we have discussed at length include the Bonferroni t, the classic multivariate procedure, the Hummel–Sligo procedure, and a combination of the classic and Hummel–Sligo procedures.

Hierarchy of Hypotheses

A research hypothesis that specifies an interaction between two or more independent variables stipulates essentially that the effect of an independent variable differs at different levels of one or more other independent variables. Thus in the event of a significant interaction effect, the focus of interest centers away from global (main) effects and onto simpler (usually cellular) units. This procedure of first testing the highest-order interaction and proceeding next to lower-order interactions and/or main effects only if the highest-order interaction is statistically nonsignificant is common to both ANOVA and MANOVA. The multiplicity of dependent variables fosters statistically significant univariate results, and these findings should be held in abeyance until overall multivariate tests indicate the appropriateness of viewing the results at that level of abstraction.

Reporting Multivariate Outcomes

Hypotheses should govern the manner of reporting results in both ANOVA and MANOVA. However, in MANOVA, the larger number and complexity of outcomes available make it desirable to have a general guide for determining what should be included in the report. The following comments are offered as a rationale for reporting MANOVA results.

1. *Test criteria.* Since MANOVA test criteria differ in sensitivity, depending on distribution of trace, it is recommended that each of the four classic criteria be reported. Their inclusion will enable the reader to evaluate hypotheses that may differ from those formulated by the investigator.

2. *Number and nature of significant roots.* The number of statistically significant roots extracted by the MANOVA approach provides a direct test of the investigator's hypothesis regarding the number of independent treatment effects detected by the set of dependent variables. The signifi-

cant roots will then require interpretation, which is most readily accomplished by reporting and interpreting the pattern of loads of the variables on the factors.

Three sets of coefficients should be reported: standardized adjusted, normalized, and correlation between each synthetic score and dependent variables. The three sets of coefficients are typically included in a MANOVA computer printout. The standardized adjusted coefficients reflect the relative importance of each of the dependent variables in contributing to synthetic score(s). Variabilities and interrelationships between variables are taken into account. The normalized (and possibly standardized) coefficients, on the other hand, are used as weights to be multiplied by values of each of the variables to form synthetic scores. These weights may be used on subsequent independent data in attempting cross-validation. The correlation coefficients between the dependent variables and the synthetic score(s) indicate the relative importance of each of the dependent variables with respect to the synthetic variable(s) and also indicate the nature and meaning of the synthetic variable(s) (Porebski, 1966).

In the MANOVA report one should also report the centroids for relevant treatment groups on each of the significant dimensions. The group centroids represent group means on the new synthetic variable(s) and permit an interpretation of treatment effects as measured by the synthetic variable(s).

To the extent that research hypotheses and MANOVA strategy focus attention on individual dependent variables, univariate means and standard deviations should be presented in tabular form, along with the appropriate univariate test statistics, such as Fisher's least significant difference (LSD), omnibus F testing, Scheffé t values, Tukey tests, and so forth.

Appendix

Hand-Calculated Example of One-Way (Simple Randomized) MANOVA

A. Data Given

Assume a purely hypothetical experiment, contrived to permit the use of small numbers and easy computation. Suppose 12 subjects are randomly assigned to three treatment conditions with the restriction that there be an equal number of subjects (4) in each group. The three treatments (independent variable) consist of three different drugs. Two outcome variables (dependent variables) are measured on each subject: heartbeat rate and galvanic skin reaction. The two dependent measures may be regarded as coded so as to permit easier hand calculation. The data are summarized in the table.

	Drug 1			Drug 2			Drug 3	
	DV_1	DV_2		DV_1	DV_2		DV_1	DV_2
S_1	2	4	S_5	6	7	S_9	7	8
S_2	1	2	S_6	4	0	S_{10}	2	2
S_3	1	9	S_7	4	6	S_{11}	0	4
S_4	8	0	S_8	7	9	S_{12}	2	8

Note: DV = dependent variable. S_i = subject i.

B. Computation of Sums-of-Squares-and-Cross-Products Matrices

The following sets of formulas that involve calculations using raw scores will be used for deriving the SSCP matrices.

Formula Set 1

$$\text{Correction term} = COR = \frac{(\Sigma X)^2}{n}$$

$$\text{Total sum of squares} = \Sigma x_T^2 = \Sigma X^2 - \frac{(\Sigma X)^2}{n}$$

$$\text{Treatment sum of squares} = \Sigma x_{\text{treat}}^2 = \frac{(\Sigma X_1)^2 + (\Sigma X_2)^2 + \dots}{n_1} - COR$$

$$\text{Error sum of squares} = \Sigma x_T^2 - \Sigma x_{\text{treat}}^2$$

Formula Set 2

$$\text{Correction term} = COR = \frac{(\Sigma X_i)\,(\Sigma X_j)}{n}$$

$$\begin{aligned}\text{Total sum of} \\ \text{cross-products}\end{aligned} = \Sigma x_i x_{jT} = \Sigma X_i X_j - \frac{(\Sigma X_i)\,(\Sigma X_j)}{n}$$

$$\begin{aligned}\text{Treatment sum of} \\ \text{cross-products}\end{aligned} = \Sigma x_i x_{j\text{treat}}$$

$$= \frac{(\Sigma X_{1_1} \Sigma X_{2_1}) + (\Sigma X_{1_2} \Sigma X_{2_2}) + \dots}{n_1} - COR$$

$$\begin{aligned}\text{Error sum of} \\ \text{cross-products}\end{aligned} = \Sigma x_i x_{jT} - \Sigma x_i x_{j\text{treat}}$$

Sum of Squares and Cross-Products within Three Treatment Cells

Computation involves each treatment unit separately, and therefore the formulas for Σx_T^2 and $\Sigma x_i x_{jT}$ are used for calculation.

Cell 1

$$\Sigma x_1^2 = 70 - \frac{(12)^2}{4} = 34 \qquad S_1^2 = 34/3 = 11.3333$$

$$\Sigma x_2^2 = 101 - \frac{(15)^2}{4} = 44.75 \qquad S_2^2 = 44.75/3 = 14.9167$$

$$\Sigma x_1 x_2 = 19 - \frac{(12)\,(15)}{4} = -26 \qquad C_{12} = -8.667$$

Cell 2

$$\Sigma x_1^2 = 117 - \frac{(21)^2}{4} = 6.75 \qquad S_1^2 = 2.25$$

$$\Sigma x_2^2 = 166 - \frac{(22)^2}{4} = 45 \qquad S_2^2 = 15$$

$$\Sigma x_1 x_2 = 129 - \frac{(21)(22)}{4} = 13.5 \qquad C_{12} = 4.5$$

Cell 3

$$\Sigma x_1^2 = 57 - \frac{(11)^2}{4} = 26.75 \qquad S_1^2 = 8.9167$$

$$\Sigma x_2^2 = 148 - \frac{(22)^2}{4} = 27 \qquad S_2^2 = 9.0000$$

$$\Sigma x_1 x_2 = 76 - \frac{(11)(22)}{4} = 15.5 \qquad C_{12} = 5.1667$$

The SSCP terms for each group are now organized into the usual matrix form in Tables A–1, A–2, A–3, and A–4.

Table A–1. SSCP Matrices for Within-Treatments Groups (Source of Error Term)

	Drug 1			Drug 2			Drug 3	
	DV_1	DV_2		DV_1	DV_2		DV_1	DV_2
DV_1	34	−26.00	DV_1	6.75	13.5	DV_1	26.75	15.5
DV_2	—	44.75	DV_2	—	45.00	DV_2	—	27.0

Table A–2. Pooled SSCP Error Matrix

	DV_1	DV_2
DV_1	67.5000	3.0000
DV_2	—	116.1667

Table A–3. Variance–Covariance Matrices for Within-Treatments Groups (Source of Error Term)

	Drug 1			Drug 2			Drug 3	
	DV_1	DV_2		DV_1	DV_2		DV_1	DV_2
DV_1	11.3333	−8.6667	DV_1	2.2500	4.5000	DV_1	8.9167	5.1667
DV_2	—	14.9167	DV_2	—	15.0000	DV_2	—	9.0000

Table A–4. Pooled Error Variance–Covariance Matrix

	DV_1	DV_2
DV_1	7.5000	.3333
DV_2	—	12.9722

Sums-of-Squares-and-Cross-Products Matrix for Treatment Effects

$$\Sigma x_1^2 = \frac{12^2 + 21^2 + 11^2}{4} - \frac{44^2}{12} = 15.16667$$

$$\Sigma x_2^2 = \frac{15^2 + 22^2 + 22^2}{4} - \frac{59^2}{12} = 8.16667$$

$$\Sigma x_1 x_2 = \frac{12(15) + 21(22) + 11(22)}{4} - \frac{44(59)}{12} = 4.66667$$

C. Tests of Homogeneity of Variance–Covariance Matrices (Error Sources)

Three different tests of homogeneity of the variance–covariance matrices (error sources) are illustrated: a generalized Bartlett test, Box's modification of the Bartlett test, and Box's F test.

The generalized Bartlett test is computed at the outset.

$\chi^2 = M = df_e \log_e |D| - \Sigma g[(n_g - 1) \log_e |D_g|]$

df_e = degrees of freedom for the error term (univariate)

$|D|$ = determinant of variance–covariance matrix pooled from within three treatment groups

$|D_g|$ = determinant for single within-treatments variance–covariance matrix

1. The determinants are calculated as follows:
   a. The appropriate matrix is factored to determine eigenvalues (λ)
   b. Successive products of eigenvalues are calculated:
   $$|D| = \pi_i \lambda_i$$
2. Factoring the pooled variance–covariance matrix from error sources yields two eigenvalues: $\lambda_1 = 12.9924$ and $\lambda_2 = 7.4798$. The determinant $|D| = (12.9924)(7.4798) = 97.180553$.

3. Factoring each of the within-treatments group variance–covariance matrices produced the results shown below.

	Drug 1	Drug 2	Drug 3
λ_1	21.9750	16.4282	14.1252
λ_2	4.2750	.8218	3.7915
$|D_g|$	93.943125	13.500694	53.555695

4. M, or χ^2, is then calculated.

$$
\begin{aligned}
M = \chi^2 &= 9 \log_e (97.180553) - 3 \log_e(93.943125) \\
&\quad + 3 \log_e(13.500694) + 3 \log_e(53.555695) \\
&= 41.18904 - 3(4.54329) + 3(2.6020564) \\
&\quad + 3(3.9806759) \\
&= 41.18904 - (13.62987 + 7.8061692 + 11.942027) \\
&= 41.18904 - 33.378066 \\
&= 7.810974
\end{aligned}
$$

Where

$$df = \frac{p(p + 1)(k - 1)}{2}$$

p = number of dependent variables
k = number of treatment groups

$$df = \frac{2(2 + 1)(3 - 1)}{2} = \frac{2(3)(2)}{2} = \frac{12}{2} = 6$$

The computed chi-square ($x^2 = 7.810974$, $df = 6$) fails to attain significance at the .05 level. Therefore the hypothesis of homogeneity of variance–covariance matrices, which are pooled to form the error term, is retained.

The use of a correction factor improves the accuracy of the Bartlett test. The correction factor (C) is found as follows:

$$C = \frac{2(p^2) + 3p - 1}{6(p + 1)(k - 1)} \left(\sum_{i = 1}^{k} \frac{1}{df_i} - \frac{1}{n - k} \right)$$

p = number of dependent variables
k = number of treatment groups
df_i = number of degrees of freedom associated with each separate treatment unit

$$C = \frac{2(2^2) + 3(2) - 1}{6(2 + 1)(3 - 1)} \left[\frac{1}{3} + \frac{1}{3} + \frac{1}{3} - \frac{1}{(12 - 3)}\right]$$

$$= \frac{2(4) + 6 - 1}{6(3)(2)} \left(1 - \frac{1}{9}\right)$$

$$= \frac{8 + 5}{18 (2)}(1 - .11111)$$

$$= \frac{13}{36}(.88889)$$

$$= .36111(.88889)$$

$$= .3209876$$

$$\chi^2 \text{ corrected} = \chi^2 (1 - C)$$
$$= 7.810974(1 - .3209876)$$
$$= 7.810974(.6790124)$$
$$= 5.304$$

The corrected chi-square value, with six degrees of freedom, also fails of significance at the .05 level.

5. Where a small number of degrees of freedom is associated with the error unit (single treatment group in the example problem) and/or the number of dependent variables and number of treatment groups exceeds six, a better approximation of the test for homogeneity of variance–covariance matrices is obtained by an F test developed by Box (1949).

$F = M/b$
M = uncorrected χ^2 value obtained earlier
$b = f_1/[1 - A_1 - (f_1/f_2)]$

$$A_1 = \left(\sum_{i = 1}^{k} \frac{1}{df_i} - \frac{1}{df_e}\right)\frac{2 p^2 + 3 p - 1}{6 (k - 1)(p + 1)}$$

$$A_2 = \left(\sum_{i = 1}^{k} \frac{1}{df_i^2} - \frac{1}{df_e^2}\right)\frac{(p - 1)(p + 2)}{6 (k - 1)}$$

If $(A_2 - A_1^2)$ is positive,

$f_1 = .5(k - 1) p(p + 1)$
$f_2 = (f_1 + 2)/(A_2 - A_1^2)$
$b = f_1/[1 - A_1 - (f_1/f_2)]$
$df_n = f_1 \qquad F = M/b$
$df_d = f_2$

If $(A_2 - A_1^2)$ is negative,

$$f_1 = .5(k - 1) p(p + 1)$$
$$f_2 = (f_1 + 2)/(A_1^2 - A_2)$$
$$b = f_2/[1 - A_1 + (2/f_2)]$$
$$df_n = f_1$$
$$df_d = f_2 \qquad F = f_2 \, M/[f_1(b - M)]$$

Consistent with the use of earlier symbolism,

$M = C$ (uncorrected)
p = number of dependent variables
k = number of treatment groups
df_e = degrees of freedom for univariate ANOVA error term
n_i = number of subjects per treatment group

For the example problem, calculation of the Box test is as follows:

$$A_1 = \left(\frac{1}{3} + \frac{1}{3} + \frac{1}{3} - \frac{1}{9}\right) \frac{2(2^2) + 3(2) - 1}{6(3 - 1)(2 + 1)}$$

$$= \left(1 - \frac{1}{9}\right) \frac{2(4) + 6 - 1}{6(2)\, 3}$$

$$= (1 - .1111111) \frac{8 + 5}{12(3)}$$

$$= .8888889 \left(\frac{13}{36}\right)$$

$$= .8888889(.3611111)$$
$$= .3209876$$

$$A_2 = \left(\frac{1}{3^2} + \frac{1}{3^2} + \frac{1}{3^2} - \frac{1}{9^2}\right) \frac{(2 - 1)(2 + 2)}{6(3 - 1)}$$

$$= \left(\frac{1}{9} + \frac{1}{9} + \frac{1}{9} - \frac{1}{81}\right) \frac{1(4)}{6(2)}$$

$$= \left(\frac{3}{9} - \frac{1}{81}\right) \frac{4}{12}$$

$$= .3209877 \,(.3333333)$$
$$= .1069958$$

$A_2 - A_1^2 = .1069958 - .3209876^2$
$ = .1069958 - .103033$
$ = +.0039628$

Since $A_2 - A_1^2$ is positive,

$f_1 = .5(3 - 1)\ 2(2 + 1)$
$ = .5(2)\ 2(3)$
$ = 1(2)\ 3$
$ = 6$

$f_2 = (6 + 2)/.0039628$
$ = 8/.0039628$
$ = 2018.7746$

$b = 6/[1 - .3209876 - (6/2018.7746)]$
$ = 6/(.6790124 - .0029721)$
$ = 6/.6760403$
$ = 8.8752105$

$F = 7.811/8.8752105$
$ = .8800918$

$df_n = f_1 = 6$
$df_d = f_2 = 2018$
$p \rangle .05$

The hypothesis of homogeneity of error sources (variance–covariance) matrices) is retained.

D. Principal-Components Analysis (Asymmetric Matrix)

First the inverse of the pooled SSCP error matrix is obtained.

	SSCP (pooled error matrix) (W)		Inverse of SSCP (pooled error matrix) (W^{-1})	
	DV_1	DV_2	DV_1	DV_2
DV_1	67.5000	3.0000	.0121659	−.0007466
DV_2	3.0000	116.7500	−.0007466	.0080511

Then the inverse matrix (W^{-1}) is postmultiplied by the SSCP treatment (B) matrix.

	W^{-1}				B	
	DV_1	DV_2			DV_1	DV_2
DV_1	.0121659	−.0007466	$\times$	DV_1	15.1667	4.6667
DV_2	−.0007466	.0080511		DV_2	4.6667	8.1667

		$W^{-1}B$	
		DV_1	DV_2
=	DV_1	.1810324	.0506774
	DV_2	.0262486	.0622668

The asymmetric matrix $W^{-1}B$ is then subjected to a principal-components analysis that extracts eigenvalues. Two eigenvalues are derived: $\lambda_1 = .2366$ and $\lambda_2 = .0548$. Each of the two eigenvalues may be tested for statistical significance using

$\chi_i^2 = \{N - 1 - [(p + k)/2]\} \ln (1 + \lambda_i)$
$df_i = p + k - 2$
χ_i^2 = chi-square value for the ith root
N = total number of subjects
p = number of dependent variables
k = number of treatment groups
λ_i = eigenvalue associated with the ith root

The first eigenvalue is subjected to the chi-square test of significance.

$\chi_1^2 = \{12 - 1 - [(2 + 3)/2]\} \ln (1 + .2366)$
$\quad = [11 - (5/2)] \ln (1.2366)$
$\quad = 8.5 \ln (1.2366)$
$\quad = 8.5(.2128)$
$\quad = 1.808$

$df_1 = 2 + 3 - 2 = 5 - 2 = 3$
$p_1 \rangle .05$

It is apparent that the first root (synthetic variable) fails to separate the treatment groups significantly. If the first root is nonsignificant, then the second must also be nonsignificant. However, the calculations are given for the purpose of illustration.

$X_2^2 = 8.5 \ln (1.0548)$
$\quad = 8.5 (.0534203)$
$\quad = .454$

$$df_2 = 2 + 3 - 2(2) = 5 - 4 = 1$$
$$p_2 \rangle .05$$

E. MANOVA Tests

The four classic MANOVA tests are derived from the eigenvalues resulting from a principal-components analysis of the $(W^{-1}B)$ matrix as just shown. The MANOVA tests are now each computed.

1. Wilks's lambda

$$\Lambda = \pi (1 + \lambda_i)^{-1}$$
$$= \frac{1}{1.2366} \cdot \frac{1}{1.0548}$$
$$= .7666561$$

The Wilks's lambda value is customarily expressed as a generalized F, or R (Rao), value.

$$F \text{ or } R = \frac{(1 - \Lambda^{1/s})/p}{\Lambda^{1/s}/(N - p - 2)}$$
$$df_n = 2p$$
$$df_d = 2 (N - p - 2)$$
s = number of roots extracted by principal-components analysis
N = total number of subjects
p = number of dependent variables
$$F = \frac{(1 - .767^{1/2})/2}{.767^{1/2}/(12 - 2 - 2)}$$
$$= \frac{(1 - .8757853)/2}{.8757853/8}$$
$$= \frac{.1242147/2}{.1094731}$$
$$= \frac{.0621073}{.1094731}$$
$$= .5673293$$

$$df_n = 4$$
$$df_d = 2 (12 - 4) = 16$$
$$p > .05$$

The F or R test on Wilks's lambda fails of significance at the .05 level.

2. Lawley–Hotelling Trace Criterion

$$\tau = \sum_{i=1}^{s} \lambda_i$$

s = number of roots extracted by principal-components analysis
τ = .2366 + .0548
    = .2914

Special tables (see Timm, 1975) are used to determine the probability value associated with given values of tau. Table entry requires the calculation of three parameters that are identical to those required for Roy's greatest characteristic root and for Pillai's trace. The calculation of the three parameters is illustrated next.

$$S = \min(df_h, p)$$
$$m = (|df_h - p| - 1)/2$$
$$n = (df_e - p - 1)/2$$

df_h = degrees of freedom associated with hypothesis under test
p = number of dependent variables
df_e = degrees of freedom associated with error term

$$S = \min(2, 2) = 2$$
$$m = (|2 - 2| - 1)/2 = (0 - 1)/2 = -1/2 = -.5$$
$$n = (9 - 2 - 1)/2 = 6/2 = 3$$

The existent Lawley–Hotelling tables do not contain values for the parameters $S = 2$, $m = -.5$, and $n = 3$ and require extreme extrapolation, along both the m and the n axes. Therefore no interpretation of probability value is made.

3. Pillai's Trace Criterion

$$V = \sum_{i=1}^{s} \theta_i$$

$$\theta_i = \frac{\lambda_i}{1 + \lambda_i}$$

$$V = \frac{.2366}{1+.2366} + \frac{.0548}{1+.0548}$$

$$= \frac{.2366}{1.2366} + \frac{.0548}{1.0548}$$

$$= .1913 + .0520$$

$$= .2433$$

See section E2 for calculation of S, m, and n values. The table for Pillai's trace criterion is entered with $S = 2$, $m = -.5$, $n = 3$. The tabled values (for $\alpha = .05$) do not include entries for $(n = 3)$; however, for $(n = 10)$ and $(n = 5)$, table entries are .357 and .567, respectively. Obviously, a significant Pillai's trace value must exceed .567. The obtained Pillai's trace of .2433 is therefore not significant.

4. Roy's Greatest Characteristic Root Criterion

$$\theta = \frac{\lambda_1}{1 + \lambda_1}$$

$$= \frac{.2366}{1.2366}$$

$$= .1913$$

A special table for Roy's greatest characteristic root does not include the parameter values for $S = 2$, $m = -.5$, and $n = 3$ (see section E2 for calculation). However, a significance test was performed earlier on the first root. Recall that a χ^2 test on the first root was not significant at the .05 level. (See section D.)

F. Other Concepts

Since the MANOVA tests were not significant at the .05 level, no further statistical analysis of the data is appropriate. However, in order to illustrate other relevant concepts, it will now be assumed that the MANOVA tests were significant. Each of two general options will now be illustrated:

1. Univariate ANOVAs of separate dependent variables.

The appropriate sums of squares for error and treatment may be gathered from the previously calculated SSCP matrices.

	$SS_{treatment}$	df_{treat}	SS_{error}	df_{error}
DV_1	15.1667	2	67.50	9
DV_2	8.1667	2	116.75	9

df_{treat} = degrees of freedom for treatment = $k - 1$
df_{error} = degrees of freedom for error = $n_T - k$
where
    k = number of treatment groups
    n_T = total number of subjects

The previous $SS_{treatment}$ and SS_{error} are converted to variance (mean square = MS) estimates in order to form F ratios ($F = MS_{treatment}/MS_{error}$).

	MS_{treat}	MS_{error}	F	Significance p
DV_1	7.5833	7.5000	1.0111	$p > .05$
DV_2	4.0833	12.9722	.3148	$p > .05$

The two ANOVAs indicate that the dependent variables considered separately do not separate the treatment groups at the .05 significance level.

2. Examination of synthetic variable

Discriminant weights for each of the dependent variables are computed as a part of the MANOVA approach. These weights, when multiplied by the original dependent variables and summed, result in a new, synthetic variable that may be of interest in its own right. The nature of the synthetic variable may be inferred by noting the (within-groups) correlations of the original dependent variables and the synthetic variable.

Table A–5. Standardized Discriminant Weights of Dependent Variables for Forming Synthetic Variables

	Synthetic variable	
	I	II
DV_1	.3498	−.1054
DV_2	.0711	.2685

Two sets of discriminant weights were computed on the example data by
the MANOVA program. The equations for converting the two dependent
variables to the two synthetic variables are as follows:

$$I' = \quad .3498\, X_1 + .0711\, X_2$$
$$II' = -.1054\, X_1 + .2685\, X_2$$

Application of the equations to the original data is shown below.

Drug 1		I'	Drug 1		II'
.3498(2) + .0711(4)	=	.9840	−.1054(2) + .2685(4)	=	.8632
.3498(1) + .0711(2)	=	.4920	−.1054(1) + .2685(2)	=	.4316
.3498(1) + .0711(9)	=	.9897	−.1054(1) + .2685(9)	=	2.3111
.3498(8) + .0711(0)	=	2.7984	−.1054(8) + .2685(0)	=	−.8432
	=	5.2641		=	2.7627
	M =	1.3160		M =	.6907

Drug 2		I'	Drug 2		II'
.3498(6) + .0711(7)	=	2.5965	−.1054(6) + .2685(7)	=	1.2471
.3498(4) + .0711(0)	=	1.3992	−.1054(4) + .2685(0)	=	−.4216
.3498(4) + .0711(6)	=	1.8258	−.1054(4) + .2685(6)	=	1.1894
.3498(7) + .0711(9)	=	3.0885	−.1054(7) + .2685(9)	=	1.6787
	=	8.9100		=	3.6936
	M =	2.2275		M =	.9234

Drug 3		I'	Drug 3		II'
.3498(7) + .0711(8)	=	3.0174	−.1054(7) + .2685(8)	=	1.4102
.3498(2) + .0711(2)	=	.8418	−.1054(2) + .2685(2)	=	.3262
.3498(0) + .0711(4)	=	.2844	−.1054(0) + .2685(4)	=	1.0740
.3498(2) + .0711(8)	=	1.2684	−.1054(2) + .2685(8)	=	1.9372
	=	5.4120		=	4.7476
	M =	1.3530		M =	1.1869

The term *centroid* designates a treatment-group average on a synthetic
variable. Treatment groups are maximally separated on the synthetic
variables. On the first synthetic variable, treatment groups rank order
from high to low as Drug 2 (2.2275), Drug 3 (1.3160), and Drug 1 (1.3160).
On Synthetic Variable II, Drug 3 has the highest centroid (1.1869), followed by Drug 2 (.9234) and Drug 1 (.6907).

3. Interpretation of Synthetic Variables.

The within-groups correlation between the original dependent variables and a synthetic variable permits us to interpret the nature of the synthetic variable. Within-groups correlations between the dependent variables and the synthetic variables were printed by the MANOVA program as shown in Table A–6. It is apparent that Synthetic Variable I is virtually synonymous with DV_1, whereas Synthetic Variable II is essentially the same as DV_2. In the calculated example, it is obvious that the synthetic variables merely confirm the primary nature of the original dependent variables in the discriminant space.

Table A–6. Correlations between Dependent Variables and Synthetic Scores

	Synthetic variable	
	I	II
DV_1	.9667	−.2561
DV_2	.2886	.9575

References

Anderson, T. W. *An introduction to multivariate statistical analysis.* New York: Wiley, 1958.

Barker, H. R., & Barker, B. M. *Behavioral sciences statistics program library* (2nd rev. ed.). University, Ala.: University of Alabama, Reproduction Services, 1977.

Barker, H. R., & Barker, B. M. Differential sensitivity to variation in trace distribution of four MANOVA test criteria. *American Statistical Association Proceedings of the Social Statistics Section,* 1979, 490–492.

Barr, A. J., Goodnight, J. H., Sall, J. P., & Heboig, J. T. *SAS user's guide.* Raleigh, N.C.: Statistical Analysis System Institute, 1979.

Berger, M. P. F. A note on the use of simultaneous test procedures. *Psychological Bulletin,* 1978, *85,* 895–897.

Bock, R. D. *Multivariate statistical methods in behavioral research.* New York: McGraw-Hill, 1975.

Box, G. E. P. A general distribution theory for a class of likelihood criteria. *Biometrika,* 1949, *36,* 317–346.

Cattell, R. B. The three basic factor analytic research designs—Their interrelations and derivatives. *Psychological Bulletin,* 1952, *49,* 499–520.

Cohen, J. Multiple regression as a general data analytic system. *Psychological Bulletin,* 1968, *70,*426–443.

Cohen, J. *Statistical power analysis for the behavioral sciences* (Rev. ed.). New York: Academic Press, 1977.

Collier, R. O., Baker, F. D., Mandeville, G. K., & Hayes, T. F. Estimates of test size for several test procedures based on conventional variance ratios in the repeated measures design. *Psychometrika,* 1967, *32,* 339–353.

Cramer, C. M., & Bock, R. D. Multivariate analysis. *Review of Educational Research,* 1966, *36,* 604–617.

Cronbach, L. J. The two disciplines of scientific psychology. *American Psychologist,* 1957, *12,* 671–684.

Dixon, W. J. *BMDP biomedical computer program.* Berkeley, Calif.: University of California Press, 1979.

Dunn, O. J. Multiple comparisons among means. *Journal of the American Statistical Association*, 1961, *56*, 52–64.

Ferguson, G. A. *Statistical analysis in psychology and education* (5th ed.). New York: McGraw-Hill, 1981.

Finn, J. *Multivariance: Univariate and multivariate analysis of variance, covariance and regression.* Chicago: National Educational Resources, 1976.

Fisher, R. A. The use of multiple measurements in taxonomic problems. *Annals of Eugenics*, 1936, *7*, 179–188.

Fisher, R. A. *The design of experiments* (6th ed.). Edinburgh & London: Oliver & Boyd, 1953.

Harris, R. J. *A primer of multivariate analysis.* New York: Academic Press, 1975.

Hinkle, D. E., Wiersma, W., & Jurs, S. G. *Applied statistics for the behavioral sciences.* Chicago: Rand McNally, 1979.

Hotelling, H. The generalization of Student's ratio. *Annals of Mathematical Statistics*, 1931, *2*, 360–378.

Hummel, T. J., & Sligo, R. J. Empirical comparisons of univariate and multivariate analysis of variance procedures. *Psychological Bulletin*, 1971, *76*, 49–57.

Huynh, H., & Feldt, L. S. Conditions under which mean square ratios in repeated measurements designs have exact *F*-distributions. *Journal of the American Statistical Association*, 1970, *65*, 1582–1589.

Ito, K. On the effect of heteroscedasticity and nonnormality upon some multivariate test procedures. In Krishnaiah, P. R. (Ed.), *Multivariate analysis II.* New York: Academic Press, 1969.

Ito, K., & Schull, W. On the robustness of the T^2 test in multivariate analysis of variance when variance–covariance matrices are not equal. *Biometrika*, 1964, *51*, 71–82.

Kenny, D. A. *Correlation and causality.* New York: Wiley, 1979.

Keppel, G. *Design and analysis: A researcher's handbook.* Englewood Cliffs, N.J.: Prentice-Hall, 1973.

Kerlinger, F. *Foundations of behavioral research* (2nd ed.). New York: Holt, Rinehart & Winston, 1973.

Kerlinger, F., & Pedhazur, E. J. *Multiple regression in behavioral research.* New York: Holt, Rinehart & Winston, 1973.

Kirk, R. E. *Experimental design: Procedures for the behavioral sciences.* Belmont, Calif.: Wadsworth, 1968.

Knapp, T. R. Canonical correlation analysis: A general parametric testing system. *Psychological Bulletin*, 1978, *85*, 410–416.

Li, C. C. *Introduction to experimental statistics.* New York: McGraw-Hill, 1964.

Lindquist, E. F. *Design and analysis of experiments.* Boston: Houghton Mifflin, 1953.

Lord, F. M. On the statistical treatment of football numbers. *American Psychologist*, 1953, *8*, 750–751.

McNemar, Q. *Psychological statistics* (2nd ed.). New York: Wiley, 1955.

McNemar, Q. *Psychological statistics* (4th ed.). New York: Wiley, 1969.

Morrison, D. F. *Multivariate statistical methods.* New York: McGraw-Hill, 1967.

Morrison, D. F. *Multivariate statistical methods* (2nd ed.). New York: McGraw-Hill, 1976.

Mowrer, O. H. *Learning theory and the symbolic processes.* New York: Wiley, 1960.

Myers, J. L. *Fundamentals of experimental design*. Boston: Allyn & Bacon, 1966.

Myers, J. L. *Fundamentals of experimental design*. (3rd ed.). Boston: Allyn & Bacon, 1979.

Nie, N., Hull, C. H., Jenkins, J. G., Steinbrenner, K., & Bent, D. H. *SPSS— Statistical package for the social sciences* (2nd ed.). New York: McGraw-Hill, 1975.

Norton, D. W. *An empirical investigation of some effects of nonnormality and heterogeneity on the F-distribution*. Unpublished doctoral dissertation, State University of Iowa, 1952.

Nunnally, J. C. The analysis of profile data. *Psychological Bulletin*, 1962, *59*, 311–319.

Olson, C. L. Comparative robustness of six tests in multivariate analysis of variance. *Journal of the American Statistical Association*, 1974, *69*, 894–908.

Olson, C. L. On choosing a test statistic in multivariate analysis of variance. *Psychological Bulletin*, 1976, *83*, 579–586.

Overall, J. E., & Spiegel, D. K. Concerning least squares analysis of experimental design. *Psychological Bulletin*, 1969, *72*, 311–322.

Pearson, E. S., & Hartley, H. O. Charts of the power function for analysis of variance tests, derived from the noncentral F-distribution. *Biometrika*, 1951, *38*, 112–130.

Perlmutter, J., & Myers, J. L. A comparison of two procedures for testing multiple contrasts. *Psychological Bulletin*, 1973, *79*, 181–184.

Pillai, K. C. S. *Statistical tables for tests of multivariate hypotheses*. Manila, Philippines: University of the Philippines, Statistical Center, 1960.

Porebski, O. R. Discriminatory and canonical analysis of technical college data. *British Journal of Mathematical and Statistical Psychology*, 1966, *19*, 215–236.

Prentice-Dunn, S., & Rogers, R. W. Effects of deindividuating situational cues and aggressive models on subjective deindividuation and aggression. *Journal of Personality and Social Psychology*, 1980, *39*, 104–113.

Ryan, T. A. Multiple comparisons in psychological research. *Psychological Bulletin*, 1959, *56*, 26–47.

Schatzoff, M. Sensitivity comparisons among tests of the general linear hypothesis. *Journal of the American Statistical Association*, 1966, *61*, 415–435.

Scheffé, H. *The analysis of variance*. New York: Wiley, 1959.

Senders, V. L. *Measurement and statistics*. New York: Oxford, 1958.

Stevens, J. Comment on Olson: Choosing a test statistic in multivariate analysis of variance. *Psychological Bulletin*, 1979, *86*, 355–360.

Stevens, J. Power of the multivariate analysis of variance tests. *Psychological Bulletin*, 1980, *88*, 728–737.

Tang, P. C. The power function of the analysis of variance tests with tables and illustrations of their use. *Statistical Research Memoirs*, 1938, *2*, 126–146.

Tatsuoka, M. M. *Multivariate analysis: Techniques for educational and psychological research*. New York: Wiley, 1971.

Timm, N. H. *Multivariate analysis with applications in education and psychology*. Monterey, Calif.: Brooks/Cole, 1975.

Tukey, J. W. *The problem of multiple comparisons*. Unpublished manuscript, Princeton University, 1953.

Walker, H. M., & Lev, J. *Statistical inference*. New York: Holt, 1953.

References

Wallis, W. A., & Roberts, H. V. *Statistics: A new approach.* New York: Free Press, 1956.
Ward, J. H., & Jennings, E. *Introduction to linear models.* Englewood Cliffs, N.J.: Prentice-Hall, 1973.
Webster, N., & Teall, F. N. *New concise Webster's dictionary.* New York: Modern Promotions, 1972.
Winer, B. J. *Statistical principles in experimental design.* New York: McGraw-Hill, 1962.
Zimbardo, P. G. The human choice: Individuation, reason, and order versus deindividuation, impulse, and chaos. In W. J. Arnold & D. Levine (Eds.), *Nebraska Symposium on Motivation* (Vol. 17). Lincoln: University of Nebraska Press, 1969.

Index

Analysis of covariance, 33
ANOVA, 10, 13, 15, 42–43, 45–49; as statistical hypothesis-testing procedure, 10; distinction from MANOVA, 42–43; interaction effects, 54; main effects, 54
Assumptions underlying ANOVA and MANOVA, 24–27; normality of distribution, 24–27; homogeneity of variance, 24–27; unequal number of observations, 25; outlier error component, 25–26

Bonferroni t method for multiple comparisons, 31, 35; advantages and disadvantages, 35–36
Box's F test, 56, 63, 75, 91

Canonical correlation, 12; weights, 12; as most encompassing statistical model, 12
Centroids, 41, 58, 63, 66, 77, 90
Checklists for MANOVA research, 3, 4, 43–44, 102–107
Compound symmetry, 97
Computer programs, 52–53
Correlation ratio, 16
Covariance, 92
Criterion variable, 9
Critical difference, 37, 41, 77
Cross-products, 16–17, 20

Determinants, 23; homogeneity of, 26; of within cell error sources, 63
Difference scores, 100–101; nonsignificant results, 100; significant results, 100–101
Discriminant function, 11, 13, 33; correspondence between df and

multiple regression, 11; test of statistical significance, 11

Eigenvalues, 18–19, 23, 26, 34; concentrated structure, 26; diffuse structure, 26
Error, 27–32, 43–44; control of, 27–28, 31, 44; Type I, 27–31; Type II, 31–32; Type G, 43–44, 47–51, 69, 79–80; Type R, 43–44, 47–51, 69; Type S, 43–44, 47–51, 69
Eta-square, 23
Extrinsic interaction, 47

Factor analysis, 17–20, 39; principal components, 39; number of factors, 53; uncorrelated factor scores, 54; roots, 57
Factor loadings, 17–20
Factors, 60–61
F_{max} test, 92
F ratio (univariate), 32, 58–59, 66, 77

Heterogeneity of dispersion matrices, 63
Homogeneity of dispersion matrices, 56, 63, 91, 98–99
Hummel–Sligo (strategy) approach to interpretation, 38–39, 59, 88, 90
Huynh and Feldt procedure, 98–99

Independent groups design, 34, 45
Interaction effects, 54, 75
Intercorrelation matrix, 19
Intrinsic interaction, 47

Law of the independent variable, 10

MANOVA: texts describing mathematics, 1, 42; minimal level of statistical expertise, 1–2; workshops on, 2–3; checklist, 3, 4, 43–44, 102; historical origins, 3, 5; conceptual theory underlying, 3, 14; decision strategies, 3, 35–41, 106; classic research designs, 3, 33–34, 45–51; applications of, 3; hand-calculated problem, 4, 108–123; coded independent variables, 13; parallels with ANOVA, 15; tests of statistical significance, 22; table parameters, 38; computer programs, 52–53, 105; principles, 53

Matched subjects design, 33–45, 48–49; exact method, 48; ordering method, 48–49

Multiple comparisons, 29–31; a priori, 29; Newman–Keuls, 29; post hoc, 29; orthogonal, 29; Scheffé technique, 30; Bonferroni method, 31; simultaneous linear contrasts, 37

Multiple correlation, 8, 13

Multiple regression, 10, 33; coding for ANOVA, 10

Multivariate analysis: in elementary statistics textbook, 6; sequential trends, 6

Multivariate era, 6

Newman–Keuls comparisons approach, 29
N-way factorial design, 48

Orthogonal comparisons, 29–30
Outlier cells, 56, 68, 91–92

Parameters S, M, N, 23, 57
Pearson r, 8, 12

Randomized blocks design, 49
Repeated measures designs, 51, 83, 96–101; univariate procedure for, 98–99; multivariate procedure, 99–101
Research designs: classic, 45–51; mixed designs, 49–51, 74–95; Lindquist Type I, 50, 74–82, 96; Lindquist Type III, 50–51, 82–95, 96; between-subjects design, 49–50, 74–95; within-subjects design, 49–50, 74–95

Scheffé technique, 30
Simple randomized design, 47, 53, 55

Simultaneous linear contrasts, 37
Statistical power, 31, 33–35, 73–74; power curves, 32; power estimation, 34, 73; for sample size estimation, 34; Pearson–Hartley power chart, 73
Subject as own control design, 33, 45, 48–49; exact method, 48; ordering method, 48
Synthetic variables, 8, 9, 18, 36, 39; composite scores, 8; factor scores, 8; discriminant scores, 8; canonical scores, 8; orthogonal scores, 8
Synthetic weights: a priori, 40; post hoc, 40

Tang power tables, 32–33
Tests of statistical significance, 22, 36–38; Wilks's lambda criterion, 22, 36–38, 54, 56–57, 84–85, 104; Lawley–Hotelling trace criterion, 22, 36–38, 54, 56–57, 84, 85, 104; Roy's greatest characteristic root criterion, 23, 36–38, 54, 56–57, 84–85, 94, 104; Pillai's trace criterion, 23, 36–38, 54, 56–57, 84–85, 104; tables for, 23, 57; differential sensitivity, 54; selection of, 104
Trace, distribution of, 24; computer simulations, 24
Treatment × Levels design, 33
Treatment × Subjects design, 33, 49–51, 70–74, 96–97; randomized blocks, 49; pretest and posttest, 50, 79; repeated measures, 51, 83, 96–101
Treatment effect: one dimension, 103; two or more dimensions, 103–104
t test, 12
t test for independent groups, 45–51
t test for matched pairs (subject as own control), 45–51, 96–97
Two-way factorial design, 45, 62, 83
Type I error, 27–31, 73; control of, 27–28
Type II error, 31–32, 73; control of, 31

Variables: synthetic, 8, 9, 20–21, 39, 59; criterion, 9; differential status, 9, 10; independent, 10, 11; dependent, 10, 39, 54; synthetic weights used to form, 39–40
Variance–covariance matrices, 56; homogeneity of, 56, 97, 99–100

Zimbardo's deindividuation theory, 82